John George Wood

Out of Doors

A Selection of Original Articles on Practical Natural History

John George Wood

Out of Doors
A Selection of Original Articles on Practical Natural History

ISBN/EAN: 9783337026639

Printed in Europe, USA, Canada, Australia, Japan

Cover: Foto ©berggeist007 / pixelio.de

More available books at **www.hansebooks.com**

STEALING A HIPPOPOTAMUS.

OUT OF DOORS:

A SELECTION OF ORIGINAL ARTICLES ON PRACTICAL NATURAL HISTORY.

BY THE

REV. J. G. WOOD, M.A., F.L.S.

AUTHOR OF 'HOMES WITHOUT HANDS' 'BIBLE ANIMALS' 'INSECTS AT HOME'
'INSECTS ABROAD' ETC.

LONDON:
LONGMANS, GREEN, AND CO.
1874.

PREFACE.

THE PRESENT VOLUME is composed of a selection of original articles on practical Natural History, which have been contributed from time to time to various periodicals, and are republished by the kind permission of the proprietors.

From *London Society* are taken ' A January Day at Regent's Park,' ' The Children of the New Forest,' ' Turkey and Oysters,' and ' Our River Harvests.' ' The Home of a Naturalist ' is from the *Cornhill Magazine*, and ' A Summer Walk through an English Lane ' and ' The Repose of Nature ' are from the *St. James's Magazine*. Five essays were written in the *Dark Blue*, namely, ' A Sand Quarry in Winter,' ' Under the Bark,' ' Mrs. Coates's Bath,' ' A Blackberry Bush in Autumn,' and ' De Monstris.' The articles on the

'Wood Ant,' the 'Green Crab,' 'Medusa and her Locks,' and 'My Toads' were contributed to *Once a Week*; and the two concluding essays, 'Life in the Ocean Wave' and 'Our Last Hippopotamus,' appeared in the *Daily Telegraph.*

The reader will probably see that the first twelve essays are arranged according to the seasons of the year, beginning with a winter of activity, and ending with a winter of repose.

CONTENTS.

LIST OF ILLUSTRATIONS.

FULL PAGE.

IN THE TEXT.

OUT OF DOORS.

A JANUARY DAY AT REGENT'S PARK.

HAVING always felt a strong interest in the economy of animated nature, I was recently led by a casual conversation to recall a visit paid to the Zoological Gardens in the coldest part of a winter now long passed away, and to reflect with some regret that the only reminiscences of that visit were a dim recollection of a polar bear paddling in some half-frozen water, and a general idea of ubiquitous straw. I therefore determined to watch for the first defined frost, and to renew my acquaintance with the gardens as soon as the temperature should be sufficiently severe for the purpose.

To the lover of all animated beings the sight could not fail to be most interesting, considering the different elements involved. Within a comparatively narrow space are assembled a variety of living creatures from all parts of the world, forming a collection at present unrivalled, and bidding fair to increase year by year. From the frozen circle of the pole to the burning belt of the equator come representatives of the fauna of every land, gathered together in the grounds of the Zoological Society like the beasts of old in the ark, though happily with more space to move, and enjoying better ventilation. Beasts, birds, reptiles, fishes, and even the lowest forms of animal life, inhabit these wonderful gardens, which contain very nearly eighteen hundred specimens to be fed and tended daily, and to be placed as nearly as possible in the same conditions which they would have occupied in their native land.

Some of these creatures inhabit the lofty mountains, while others pass an almost subterranean life in the plains and valleys ; some require a warm and moist atmosphere, while others would die unless they could breathe a cold and dry air ; one must live almost wholly in water, while another would be injured even by a momentary immersion therein. Some animals, again, are fierce, savage, and powerful, requiring heavy iron bars and resolute keepers, while others are so soft and gentle in their nature that they require to be tended as carefully and watchfully as infants. Some are sullen and morose, others are affectionate and cheerful ; some are shy, others are familiar ; and, in short, there is hardly a mental phase that does not find a representative in the creatures forming this collection.

In the matter of food, again, there is as great a diversity as in climate or disposition.

The carnivora, whether furred, feathered, or scaled, of course require animal food, which, again, is varied to suit the particular species that need it: the lions and their kin eating flesh meat; the seals and others needing fish; and the snakes requiring living prey, such as frogs, birds, rabbits, and similar creatures. As to the variety of vegetable food which is needed to meet the wants of the beasts and birds that live on herbs, leaves, and seeds, it is too complicated for any detailed account. Add to all these elements the individual idiosyncrasies of many valuable specimens, and some idea may be formed of the labour involved in keeping such an establishment in proper order.

Few persons have the least notion of the intellect, perseverance, and watchfulness that are daily exercised in this place, of the ready invention required to meet sudden and unexpected difficulties, and the resolute courage by which alone they can be overcome. Few of the visitors who stroll leisurely from cage to cage think of the exceeding benefit conferred on science by this collection, and the valuable additions to zoological knowledge that have been made through its means.

Many curious and disputed points in animal physiology have been cleared up, which otherwise must have been left to conjecture and theory, and the pains taken about the needful experiments are as surprising as they are generally unknown. In order to ascertain but a single mooted point, a staff of observers has been organised, relieving each other at regular intervals, never quitting their posts for a single instant, either day or night, and keeping their ceaseless watch lest at some unguarded moment the golden opportunity might be lost, perhaps never to recur. Any one who wishes to form an idea of the accuracy, perseverance, and watchfulness that are exercised on such occasions, need but refer to the celebrated experiments conducted by Professor Owen, in order to settle certain difficulties in the development of the kangaroo.

In spite of all the care lavished upon this institution, winter is always an anxious period. Bearing, therefore, all these and many other considerations in my mind, it was with no small interest that I entered the Zoological Gardens on Old Twelfth Day, Saturday January 18th, 1862, the thermometer then indicating a temperature of 24° Fahr., and a tolerably sharp breeze blowing.

On casting a comprehensive glance at the various enclosures, the first object that caught my eye was a creature something like a grenadier's cap, or a lady's muff set on end, reared against the bars of the enclosure, and gently swaying its body backwards and forwards. Presently it began to sidle along the bars, still standing or sitting upright, and being rendered so indefinite in shape, by intervening twigs, wires, and posts, that I could not make it out at all. However, it soon turned its odd, wise-visaged head, and all the Beaver sat confessed. As the beaver is a North-American animal, accustomed to brave the terrible winters of that climate, and quite familiar with ice, I should not have troubled myself about it, but for its movements and general demeanour partaking so largely of the absurd, and its perfect contentment amid conditions that would seem the very acme of discomfort to a human being. After watching the inquisitive creature for some time, it was easy to appreciate the veneration in which its intellectual powers are, or were once, held by the noble savage of North America, who would naturally reverence an animal that could build a house far superior to his wigwam, and was clever enough to dam up a too shallow stream, and to lay by a store of food for the winter— two branches of social economy which the savage mind would not have conceived, far less have executed.

Dripping with water which froze almost immediately on touching the ground, and had already covered the enclosure with spots and paths of ice, the beaver looked as luxuriantly comfortable as a cat on a hearthrug, and was enjoying himself amazingly. Sometimes he would patter round his pond, his flat tail dragging behind him ; then he would make for the water, flounce into the halffrozen liquid with a splash that caused the nerves to shudder in misplaced sympathy, make a great turmoil with paws and tail, and then emerge, walk to the bars with the water dropping from every hair, seat himself on end, holding with his feet to the iron fence, and, with a calmly inquisitive air, inspect the carriages passing on the road, or the visitors who happened to approach his home.

Good store of tree trunks and branches have been considerately furnished to him, and the grooves on the wood, and the chips which strew the enclosure, are convincing proofs that the kindness of his attendants is not wasted, and that his teeth have been rightly exercised.

Near this animal is another of the same species, not so large, and inhabiting quite a little enclosure with a mere trough of water, transformed by the united exertions of the animal and the frost into an unpleasing compound of water, mud, ice, and chips. The animal was mightily hard at work when I came to its cage, carrying a bundle of straw in its mouth for some time, washing it well, and then rearing the bundle carefully against the angle of its den, and tucking it down neatly with its paws. I thought it was playing at building a dam.

It was evident that as far as the beaver was concerned there was no cause for anxiety, and I therefore passed on to see how the inhabitants of Southern Africa were comporting themselves under the present circumstances.

As usual, the hippopotamus was enjoying himself in his bath, rolling about and wallowing in the familiar element in a lazily contented fashion, ever and anon slowly submerging the whole of his unwieldy person below the surface, with that remarkable power of adaptability which permits such animals as the hippopotamus and elephant to rise and sink at will, thus making themselves heavier or lighter than an equal bulk of water without needing to expel or inspire air.

This is a most interesting performance, especially to a practical swimmer, and is probably achieved by compressing the muscles of the chest so as to reduce the bulk when the creature desires to sink, and allowing itself to expand to its former dimensions when it wants to rise.

The native habits of this huge animal are well exhibited in the magnificent male specimen now in the gardens, and it is curious to see how wonderfully the creature is fitted for an aquatic existence. Heavy, corpulent, and unwieldy as it appears on land, its legs set so widely apart that when it walks in high grass the limbs of each side make a separate path, leaving a ridge of untrodden grass between them, it assumes quite another aspect as soon as it enters the water, and, in the easy playfulness and almost grace of its movements, affords as great a contrast to its former clumsiness as does the swan proudly sailing on the lake to the same bird uncouthly waddling on the shore.

As the tank in the enclosure was so thickly covered with ice that the animal might have practised sliding, but would have found swimming next to impossible, the hippopotamus was forced to content himself with the small tank within his house, where the water is kept at a moderate temperature by artificial means, and the atmosphere is such as this delicate though monstrous animal can breathe with safety. The attendants are peculiarly careful of so valuable a creature, and have made arrangements for cleansing its house without sending their charge into the outer air during the operation.

The giraffes are nearly, if not quite, as delicate as the hippopotamus; and are obliged to content themselves with gratifying their very inquisitive natures by inspecting the visitors who occasionally pass through their warm house, and would like to feed the graceful and gentle creatures, were not all such attempts sternly prohibited by the watchful guardians. It is rather remarkable that within a yard or two of each other are located specimens of animals which inhabit the same land, and yet are as strongly contrasted in shape and habit as if they came from opposite portions of the globe.

The elands are well and comfortable, and appear to be tamer than was the case a few months ago. They are able to withstand the weather better than the hippopotamus and the giraffe, being, indeed, mighty mountain climbers in their native land, and therefore accustomed to a low temperature. I may here mention that the healthy condition of these magnificent antelopes, and the comparative ease with which they are bred in this country, afford most gratifying encouragement to the efforts now being made in many quarters to acclimatize in our own land the useful and ornamental inhabitants of other parts of the world, and show in a striking manner the national value of a collection upon which so much time is spent, and to which such stores of knowledge are cheerfully dedicated.

The acquisition of a single new article of food, whether animal or vegetable, is no slight boon to a country, and it is almost impossible to exaggerate the benefits that will accrue to this land if we can fairly establish this splendid antelope as a denizen of our parks or paddocks. When adult and well fed it is as large as a prize ox; its meat is of a peculiarly delicate and piquant flavour; its fat, a handbreadth thick, is thought to surpass that of venison, while the marrow is of such transcendant merit that a South African hunter can hardly trust himself to think about it. There are, of course, many difficulties in the way, inasmuch as the animal has not yet become civilised, and is apt to display an amount of irascibility that is rather terrifying in an animal that wears horns as sharp and powerful as those of an Andalusian bull, that can leap a fence or chasm from which the boldest hunter would recoil, and can charge down a precipitous hill with the speed and sure foot of the chamois. Still it is possible that in successive generations this evil temper may be elimated by careful management; and it is to be hoped that before the lapse of many years the eland may be as common in our parks as the fallow deer.

Nor is this the only creature which is being bred at the Zoological Gardens with the intention of acclimatizing it. Among quadrupeds the bison of North America and the kangaroo of Australia are among the number of the intended denizens of this country, while among the birds may be noticed a great number of species belonging to the poultry and the pigeons, such as the splendid curassows of tropical America, and the large wonga-wonga pigeon of Australia. France and England are uniting in the same great object by means of their respective Societies of Acclimatization, and should Europe be hereafter enriched with the valuable beasts and birds that are now being gradually accustomed to the conditions of a strange land, it is to be hoped that posterity will not forget how deep a debt of gratitude they owe to the Zoological Gardens of London, the property of a private Society.

Desirous of seeing how the cold weather was borne by the ostriches, I went to look at my old friends, whom I found shut up in their houses, but very glad to see me, and as desirous as ever of eating any object they could snap up. The shining top of my pencil-case was a wonderful object to these inquisitive and voracious birds, and it was most absurd to see all the heads bobbing up and down, the large brown eyes gleaming with excitement, and the wide mouths opened and shut with impatience, just because I was writing with a pencil that had a glittering top.

The temperature was 45° Fahr. in this department, and the ostriches and cassowaries were quite at their ease, as probably was the apteryx; but as the latter bird was hidden, as usual, behind her bundle of straw, and was in all likelihood fast asleep, her exact condition could not be ascertained. There are plenty of odd birds in these gardens, but the apteryx without doubt is the oddest of all existing feathered bipeds. ·Wingless, tailless, thick-legged, long-beaked, and brown-coated, she is about as queer a specimen of a bird as can well be imagined ; and, as a climax to her eccentricities of behaviour, persists, though a spinster apteryx living in more than conventual celibacy, in laying enormous eggs, each of which weighs one-fourth as much as the parent bird. Several emus, however, were trotting about in the open air, and were pecking here and there at the grass, or poking their long necks over the rails of the enclosure, as gaily as in the summer months, though the ground was frozen to a strong hardness, firm ice was at their feet, and the sounds of boys sliding were heard just outside the fence.

There are, of course, far too many beasts and birds in this collection to be separately examined, so I turned my steps towards the tunnel, walking casually through the parrot house, and dropping a word or two of recognition to my garrulous acquaintances, and then passing out to pay a visit to the piping crows of Australia, who were chattering away in the open air, brisk and saucy as ever, and always ready for a conversation. One of them, the white-backed species, was singularly lavish of his conversational powers, and engaged in a contest of strength on the spot. First the bird whistled a few wild notes, and then paused, while I did the same. Twisting his head on one side, and looking up knowingly with one eye, he waited for my lead, and imitated my whistle with wonderful fidelity. He got quite excited at last, flew to his perch, thence to the wires on a level with my face, clung firmly with his strong claws, poked his beak through the interstices of the intersections, and fairly screamed with exultation. Meanwhile his companion was making the best of his time by pecking my boots.

Pleasant as this amusement was, the hours were passing, and the wind was chilly, so I bade farewell to the piping crow, and cruelly left him, in spite of his repeated attempts to recall me by screams and whistles.

Mag, in the next compartment, was cheerful enough; so were the ravens, with whom I exchanged a friendly croak in passing, and allowed them their usual bite at my pencil.

The elephant and the rhinoceros have been too long residents to care much for the vicissitudes of an English climate. The former was swinging itself from side to side in his den with that peculiar movement which seems instinctive to the creature, and may possibly answer as a succedaneum for walking exercise. The latter was serenely munching a truss or so of straw, his nose in the air, slapping his lips together with every sidelong movement of his mouth, while from his big lungs issued an occasional grunt of satisfaction, though certainly the substance which he was eating seemed absurdly incapable of affording any nourishment to the system, or gratification of the palate. None

of these animals are allowed to expose themselves to
the virulence of so frosty and inclement a day.

The reptile house is always kept at so uniform a
temperature that winter's cold or summer's heat makes
hardly any perceptible difference. The fine specimen
of the North African monitor was in a state of great
excitement, endeavouring apparently to climb up the
plate-glass front of his cage, and ever and anon falling
back ignominiously, only to resume the attempt with
renewed vigour. It was astonishing what a noise the
creature made by scratching his claws and rubbing his
chin against the glass, and to what unexpected attitudes
his lithesome body and slender neck could be writhed.
The reptile was shedding its epidermis, which hung in
shreds and patches from different parts of the body,
showing the bright scales beneath as they were freed
from their effete covering. The creature was very persevering in his exercise, continually darting out his
long and deeply-cleft tongue, looking, indeed, as if it
had been furnished by nature with two slender pointed
tongues, and affording an admirable opportunity for
studying the arrangement of the beautiful spotted
scales on the lower surface of its body.

Its near neighbour, the rock snake, or pythoness,
as it is just now the fashion to call her, was not visible,
being, in fact, as well as could be expected under the
circumstances, and lying under her blanket, coiled like
a shallow cone around her new-born family of eighty or
ninety eggs. The chameleons were perched immoveably

as usual on the branches with which the cage is plentifully furnished, and gave no signs of life, except occasionally turning one great green-pea of an eye upwards or downwards, as the case might be. The African cobra lay flat upon the floor of its cage, but on seeing a human face, surmounted by a hat, coming close to the glass, it became rapidly excited, spread its hood, puffed out its body, and raised itself as if threatening an attack. Not wishing to be the cause of a possible injury to a valuable reptile by letting it strike its nose against the glass, as it was evidently preparing to do, I passed on to the bull frogs, and so out of the room.

In the next apartment the creatures were all doing well. A single specimen of the flying fox survives, though the keeper expressed himself as rather anxious concerning its chance of getting through the winter. That singular creature, the gigantic salamander, lay impassive as usual along the bottom of its tank, and though so remarkable an animal, attracts but little notice from visitors. Hundreds pass through the room daily without seeing it at all; and of those who condescend to cast a glance at it, the greater number express themselves as sadly disappointed. The general public has heard great tales of salamanders, and through the medium of a weighty culinary instrument bearing the same title has learned to connect the name with fire and glowing metal. Reading the name of gigantic salamander, they enter the room in a rather nervous and uneasy state of mind, expecting to see nothing less than fourteen or fifteen feet long, and hoping that the bars are strong enough to prevent it breaking from prison. Great, therefore, is their disappointment on being shown a glass tank of water such as they see in any naturalist's window, and are referred to a creature like a big black tadpole which lies grovelling quietly in one corner. Some decline to believe that the animal is the dreadful creature which they have been led to expect, and others openly aver that the whole affair is a delusion, and akin to Barnum's mermaid. Yet the beast is a wonderful beast after all, and in the eyes of naturalists is a very gigantic salamander. For, in sooth, the eft or newt is a salamander, and an eft of thirty inches in length is gigantic beyond doubt. Besides, it is very rare even in Japan, whence it comes, and its habits and general economy are very remarkable.

Nearly opposite to this salamander is a creature of unpretending form and dimensions, but still more curious in its structure and habits than even its black, flat-headed neighbour : this is the lepidosiren or mudfish of Africa, remarkable for having long been an object of contention among naturalists. Is it a fish or is it one of the frog tribe ? No one exactly knows, and, to judge from the opposite opinions expressed by the most accomplished naturalists and dissectors, no one is likely to know. Perhaps it is neither, but represents an intermediate class between the fish and the reptiles, with the heart of one and the gills of the other. This specimen has lived for about three years in the tank which it now occupies, and has grown, though slightly, in that time; thus affording a singular contrast to the specimen at the Crystal Palace, which attained a length of nearly a yard in the same time, though not nearly so large when first brought to England. But then the Crystal Palace animal got into the large hot-water basin and there lived a despotic life, feeding *ad libitum* on gold fish until he was captured and his depredations stopped, and on frogs afterwards. Should the reader pay a visit to the Zoological Gardens, as I trust soon will be the case, let him look well at the mud-fish, the Gordian knot of systematic zoology.

On my way to the lions I looked in at the wombat's cage, and there saw to my surprise that the animal, though a native of Australia, was lying curled up in one corner of the enclosure fast asleep, with the thermometer marking eight degrees below freezing point, and the wind blowing in keen and cutting blasts. The bars of the enclosure being open and of iron afforded no protection whatever, but would rather have the effect of chilling a creature that was pressed against them. The seals were naturally indifferent to the cold, and darted about in the water, or flounced their way over the rim of their bath, as if enjoying the icy coldness of their home. They ran some very good races after fish, driving up the water before them like the bows of two fast steam-boats, and had quite a struggle for the last fish. The otter, too, cared nothing about the temperature of the water, but sat on a heap of wet straw, eating his dinner, with the end of his tail in the water, and the freezing drops glittering around him. To the shivering observer, whose chilled fingers could scarcely hold the pencil, and whose heart yearned for a seat in a warm room and a large cup of hot tea, the choice of locality seemed singularly unfortunate. There, however, sat the animal, thoroughly contented with his position, holding his flounder tightly between his paws, and crunching and tugging with hearty goodwill.

The lions, tigers, and other large carnivora, are carefully defended from the outer cold by means of thick screens rigged from the eaves of the projecting roofs to the bars beyond which visitors are requested not to pass. As, however, the greater number of visitors would be sadly disappointed if they had to go away without seeing these beautiful animals, they are admitted for the nonce into the space between the bars and the cages; and in order to prevent the fierce beasts from thrusting out a paw and inflicting a wound, either in sport or anger, a strong wire grating is affixed to the front of the cage, which effectually prevents any such mishap. Notwithstanding all these precautions, and an assured conviction of the absolute security attained, I could not help instinctively starting back when the lion took it into his illogical head that I was going to steal his meat, and flew at me with flaming eyes and a roar that shook the place. I had much respect afterwards for the steady nerve of those who can endure such a charge with a firm hand and un-winking eye, and very much less contempt for the native attendants who in such cases always throw away their guns and run for their lives. The whole of these dens are kept at a comfortable temperature by hot pipes, and the animals seem as contented as in the hot summer time.

Two lions, however, in neighbouring cages became angry with each other, or perhaps jealous ; and putting their mouths to the floor just by the wooden partition, began to roar against each other to the utmost of their power. It was a grand exhibition, and would alone have been worth the trouble of the visit. The threatening sounds seemed to reverberate through every nerve, the whole building trembled as if shaken by rolling thunder, and the rest of the beasts sank into respectful silence while the kings of the forest lifted their mighty voices. No wonder that at the sound of the lion's roar the beasts of burden break their halters and flee in terror over the plain; but it is a wonder that the ostrich, the meekest looking of birds, should roar so exactly like the lion that even the native hunter cannot always distinguish the one from the other.

As if intended to produce a striking contrast to the lions, tigers, and leopards existing in a temporary hot-house, and sheltered from the chilling blasts by a screen erected expressly for the purpose, the polar bears live within ten yards of these heated localities, rejoicing in the cold, and probably thinking of the ice-fields and freezing waters of their proper home. This is one of the few northern animals whose fur retains its white hue throughout its life, experiencing no change in winter or summer. The coat of the ermine and the arctic fox alters from its dark summer tints to its snowy winter hue; not, I imagine, to aid in concealment by assimilating the colour of the animal with that of the ground, but because the pure white hue is endowed with some wondrous power of resisting the effects of cold.

I wonder whether polar bears when wild are in the habit of taking exercise in the fashion in which these specimens indulge? Do they always walk forward for six paces, and retire backwards over precisely the same ground, with as much accuracy as if they had been volunteer riflemen practising the back-step? It can hardly be too troublesome for them to turn round, and they have ample room for the purpose, being able if they choose to indulge in quite a promenade, unrestricted by the narrow limits in which those unfortunate lions and tigers are confined.

I pity those active and restless creatures with all my heart. I wish they had more appropriate residences, and am sure that if they were only permitted to exercise their limbs as intended by their Maker, they would be healthier, live longer, and display their wonderful powers in a more perfect manner. There are, of course, some difficulties attendant upon the construction of an enclosure sufficiently large to give ample room to the agile limbs of the feline race, sufficiently strong to withstand the fiercest assault of the lion, and properly roofed so as to counteract the danger of a leopard or jaguar climbing over its walls. I cannot but think, however, that it would be quite practicable to construct an enclosure that would comply with all these requisitions, and at no very great outlay of space or money. The enclosure might be common to all the feline race, and each species might be allowed to exercise in it in regular rotation. There would be no difficulty in decoying them back to their dens, as a piece of meat would effectually accomplish that design, and allow of the door of communication being closed while the animals were engaged upon their food.

The interior of the enclosure should be furnished with artificial trees, and I have often pictured to myself the magnificent sight of a pair of lions or tigers careering round their pleasure ground, exulting in their strength, or a company of leopards disporting among the branches, and displaying their lithe forms in all their spotted beauty. Look, for example, at the monkeys, and think how much we should have lost by cooping them up in little boxes, where they could hardly move, instead of giving them spacious apartments, fitted with ropes, bars, and boughs, so as to enable them to display their marvellous agility to our wondering eyes. Sure am I that a lion, tiger, or leopard, when permitted to range freely over an ample space, would present as great a contrast to the same creature uneasily dcambulating its narrow den, with its head close to the bars, and its paws slipping over the smooth wet boards, as does a monkey in a box to the same animal in a spacious apartment, or a caged squirrel to a scuggy in his native woods.

Both species of camel—the dromedary and the double-humped camel of Bactria—were quite at their ease about the weather. The former animal was standing partially in its shed, with its long neck and meek-looking head peering out at the landscape, while the latter was quietly walking about its enclosure, though the ground must have been very uncomfortable to its feet, and the water in its trough had been frozen so hard that the attendant had been obliged to break the ice, in order to allow the animal to drink.

The coypu rat seemed rather unwilling to face the cold, though attracted by a large carrot that the keeper had placed within its den. This odd, blunt-nosed, orange-toothed quadruped only emerged at intervals, ate a piece of carrot, and then returned to its warm home. I remarked that the mice are very fond of the coypu's house, and run in and out of the straw with amusing impudence. The creature evidently dislikes the ice, trying in vain to get its usual bath, and feeling sadly disappointed at finding itself arrested by the icy covering of its little pool. The reader is hereby advised to pull up a little tuft of grass by the roots, and place it in the coypu's cage, for he cannot fail to be amused by the clever and systematic manner in which the ingenious and cleanly animal picks up the grass, takes it to the water, and washes it carefully before it will condescend to nibble a single blade.

The honey-ratel, with his dark waistcoat and grey coat, was in great force, running about his cage in quite an excited fashion, and even climbing up the wires as if to survey the prospect. In the summer time of the year this animal has a habit of running continually about its den in an oval-shaped course, which is marked by the continual tread of the feet like the sawdust in a circus. The oddest part of the performance is that whenever it reaches either extremity of its course it puts its head to the ground, turns a somersault, and recommences its race. The fine specimen of that very fierce animal, called from its evil temper the Tasmanian devil, was occasionally to be seen in the open air, but it preferred the warm retreat of its straw-sheltered shed.

The winter aviary, which is ingeniously constructed so as to admit of glazed casements in addition to the wires, is employed as the home of several valuable and delicately constituted animals. In the central compartment is a remarkably fine specimen of that curious animal popularly called the Tasmanian wolf, but which really is not a wolf at all, but one of the marsupial tribe, related to the opossum and their kin. The beautiful çariamas thrive well ; and as they sat on their perch with bent knees, and head sunk so deeply upon the breast that the curious feathery crest that decorates the head was scarcely perceptible, they could hardly be re-cognised as the same birds who stalk about their cages with long and haughty strides, erect gait, and bold, intelligent gaze. Perhaps, however, the most curious inhabitants of this aviary are the crested eagles, fine, handsome birds, notable for an erect tuft or plume of black feathers upon their heads, not unlike the ostrich plumes of a lady's court dress.

The last animals visited were our volatile friends the monkeys, who seemed none the worse for the com-paratively close quarters to which they are confined in severe weather. The house is rather dark just now, because the windows are thickly banked up with straw, a precaution necessary lest the monkeys should be chilled by coming in contact with the cold glass. The temperature of the room is very comfortable, but not unpleasantly warm, and is maintained by a partly open stove or fireplace in the centre. I was sorry to miss my dear old friend Sally, the spider-monkey, whose gentle manners and wonderful length of limb I have often admired. Agile as are all the monkey tribe, Sally was certainly the most active I have yet seen in this country, and her performances on the rope would have put the combined efforts of a dozen Leotards or Blondins to shame. I shall never forget her happiness when dancing and swinging about on a clothes line in a garden near Reading, the curious air with which she contemplated the surrounding objects, and the look of piteous entreaty with which she deprecated the order to leave her rope and return to her seat on the back of a chair near the kitchen fire.

The funny little Capucin monkey was as amusing as ever with his nuts and pebble, using the latter in the light of a hammer and smashing the nutshells with wonderful certainty. The odd little creature has a perfect passion for hammering, and had battered the woodwork of his cage so severely that the keeper was forced to take away the stone, and now lends it only when it is wanted. Even the hard, angular shell of the Brazil nut is broken by this clever little animal, and the keeper told me that he—the monkey, to wit—could hardly have a greater treat than to be given a hammer and a board with a nail partly driven, so that he might take the hammer and finish driving the nail.

The great anubis baboon sat sulky and impassive on his perch, his chin sunk on his breast, his limbs gathered up into marvellously small compass, and his toes holding tightly to the bars. Offerings of nuts and other dainties failed to propitiate his frigid dignity ; and it was not until the keeper spoke to him that he would condescend to notice the gifts that were freely proffered. Even after taking the nuts and pieces of cake, he just put them in his mouth, ascended again to his perch, and resumed his former misanthropical attitude. Large store of straw is placed in his cage, and when evening approaches he retires to the farther corner of the cage, creeps into the heap of straw, and with hands and feet disposes it around him in such a manner that not a vestige of his person can be seen.

In a large cage, where a number of the smaller monkeys are congregated, the ruling power of the establishment was evidently the huge white and black cat, who lay calmly dozing among all the restless quadrumana, supremely indifferent to their noisy gambols. Even when a graceless monkey leaped on her back from a perch, and was straightway assaulted by one of his companions, the cat did not even open her eyes, but lay purring, with her paws tucked comfortably under her chin, in utter unconcern. Pussy has been used to monkeys for so long a time that she is quite uncomfortable out of their presence, and cannot endure being placed in the open air. The keeper fetched her out of the cage to enable us to judge of her weight, which is really wonderful for a cat of the gentler sex, and hardly was she fairly on the ground, and the door of the cage opened, than she leapt through the aperture and resumed her former position.

No sooner did the shades of evening become perceptible than the monkeys made arrangements for the night, ceasing from their sports, and even allowing the armadillo to run about the cage according to its pleasure, without jumping on its back for a ride, or trying to pull it over as it trotted past them. They congregate together in compact bodies, presenting a most absurd effect of parti-coloured fur, intertwined limbs, and long dangling tails, and were continually struggling for the snuggest and warmest spot, which was, of course, the centre of the group. One individual was totally excluded, but he took the matter in a philosophical light, going carefully over the cage and picking up all the little bits of biscuit and stray nuts which his companions had relinquished when battling for a place on the perch.

Throughout the whole of the visit it was pleasant to denote the demeanour of the attendants, upon whose sympathetic kindness depends so much of the comfort and happiness of the animals under their charge, and the manner in which they accommodate themselves to the individual idiosyncrasies of their charges. Should the animal happen to be docile and intelligent, no one is more proud than the keeper, and no visitor can be more interested in seeing the clever performances of any creature than is the keeper in exhibiting them. It was pleasant, for example, to see the two splendid chetahs' behaviour towards their attendant, and ludicrous enough to watch him coolly sweep either individual out of his way with the broom if they happened to interfere with his movements while cleaning their cage. If they had been a pair of three months' old kittens there could not have been more confidence on the one side or playfulness on the other. As the keeper left the cage, the gentle and beautiful creatures pressed after him, but were gently put back with one hand while he took down some meat with the other. Even under such exciting circumstances, with their dinners in their sight, they displayed none of the fierce eagerness so common among the feline race when they see or smell their food, and they took the meat with even less haste than my own pet cat exhibits when the food is to his taste, and he happens to feel hungry.

Should, however, the animal be of a vicious and impracticable disposition, the keeper only seems to be amused at the various exhibitions of cross-grained temper, and laughs good-humouredly at every growl, or attempted assault.

Perhaps the reader may have remarked in the course of this slight sketch of a very wide subject, the apparent absence of all rule regarding the capability of any animal to resist the effects of cold weather and a strange climate. It is easy enough to understand that the beaver and the polar bear could be quite happy on a frosty day, and that the lions, tigers, and leopards would need protection against the chilling atmosphere. But it was hardly to be expected that the camel, which is essentially the 'ship of the desert,' made to endure long thirst and to pace for weeks over the burning sand, should walk about quite at its ease upon frozen soil, and drink from a trough in which the ice was thickly gathered. This phenomenon will perhaps give some idea of the difficulties attendant upon acclimatizing the denizen of a strange soil, inasmuch as it is quite impossible to treat one animal on a system derived from the management of another species from the same country and with similar habits. Each new species must be learned by means of repeated and cautious experiments, and to the minds of thoughtful lovers of nature, and observers of animal life, this very want of uniformity affords a better hope of ultimate success than if it were possible to reduce the management of foreign animals to a rigid system, and treat all creatures of kindred forms and similar countries on the same stereotyped principles.

THE end of November, 1872, 6.30 A.M.   Wet, wet, wet! Thermometer 34 deg.   A fierce wind blowing somewhere from the northwards, howling and shrieking through the trees, and, as can be seen even at that hour, tearing off the leaves that still keep their hold on the branches, whirling them high in air, and mixing them with the already fallen leaves which have been swept up from the ground, and tower upwards in spiral eddies before they again drop to the earth.   No moon: the sun is not due yet, but he is trying hard to drive a few pale, watery beams through the dull, leaden, black-patched canopy which does duty for a sky; and, as the eye becomes more accustomed to the semi-darkness, a few large snow-flakes are seen here and there amid all the flying leaves.   The sash is opened for a better glance of the sky, and in rushes the triumphant wind, sending all my papers flying helter-skelter about the room, and causing great confusion among the multitudinous savage weapons, implements, and ornaments with which the walls are covered.

The house is situated on the top of the hill, so that

the wind does pretty well as it likes, especially at this time of year when the foliage is off the protecting belt of trees in front, and nothing is left but their bare branches. Clearly, this is a day for home-work, for avoidance of the elements, and for cheerful fires in defiance of the colliers.

A letter from the editor, urgently requesting an article at once, because the magazine has to go to press so early in November. I had looked forward to a nice bright December day for this task—one of the many wintry days when the sun shines clearly, though coldly; when the sky is blue, when scarcely but the slightest breeze is perceptible, and when the exhilarating, bracing atmosphere almost takes away the sense of cold. Moreover, I had intended to write an article entitled, ' Under the Ground,' as a companion to ' Under the Bark,' but the perpetual rains of the last two or three months have rendered such a task all but useless. There are hundreds of insects which pass their wintre time some few inches below the surface of the earth, and I had thought of taking a limited area, digging et carefully, and jotting down the results; but there is nothing which does so much damage to most insects as wet. Cold they can bear well enough, provided they are not exposed directly to the elements, but wet is more than they can endure, and fairly drowns them, and it is for that reason that insects are often so rare after a very wet autumn.

Surely no one could be expected to go out in such

weather and dig for insects ; and if he were rash enough to do so, the chances are that no sooner did he uncover an insect than it would be blown far out of his reach. At last I bethought myself of a small, sheltered sand quarry, about half a mile from my house, and, taking with me the old familiar butcher's knife in its sheath, and some boxes, I started for the quarry.

When I visited the place in July last, it was a most lovely little spot, clothed with abundant verdure, rich in the sweet flowers of glorious summer, and musical with the twitter of joyous birds and the hum of many insects. The sky was serene and calm, with a few white clouds drifting slowly across its azure expanse, and sending their shadows travelling over the plain below. The Thames ran, a meandering blue streak, glittering here and there as the sunbeams glanced on its ripples, and bearing many a white sail and swift steamer through the valley over which it had once spread itself like a shallow lake until restrained within its limits by the mighty ' river-wall,' on which the seaweed dangles in black and green clusters.

Now, how changed is all the scene ! The quarry itself is tolerably sheltered, but above our heads the wind tears its way through the wood, and speeds over the country as if it meant to twist every tree up by the roots. Every now and then, as some fiercer gust passes along, a loud ruffling sound is heard, accompanied by a pattering as of hail, among the withered leaves that strew the ground. At first, indeed, I took it to be really hail, but presently found that it was caused by the little hard seeds of the broom, which clung somewhat loosely to their opened and twisted pods, and were shaken out by the wind. All the broom trees above had lost their seeds long ago, but these still survived in that partly sheltered spot. The rustling sound was produced by a young sycamore tree. All the leaves had been blown off it except one large leaf at the end of each twig. These clung pertinaciously to their hold, and the noise which they made was really wonderful.

No longer bright and glittering, the Thames, a dull grey stream, reflected the dull grey and leaden sky, through which no ray of sunshine could pass, and over which the black snow clouds sped with ominous rapidity. Not a sail visible, and only an occasional empty screw-collier, very much down at the stern with the weight of her engines, and her 'nose tip-tilted' as if disgusted with things in general. Far away on either side lie the marshes as they are still called—'the meshes' according to aboriginal pronunciation—and on the left is the identical 'mesh' where Pip encountered his grateful convict, and nearly met his death in the hut by the lime kiln. Not many years ago the bittern haunted these marshes, but its weird, booming cry has never been heard since the marshes were drained and cultivated. For all that may be seen now the bittern might yet be there, and a more forlorn-looking place can hardly be imagined than that dim, misty expanse

THE SAND QUARRY.

with the river winding through it, and closed in the distance by the black, tree-topped Essex hills.

As to the wind, it seems scarcely to have made up its mind what to do, or by what name. it was to be called, whether Boreas, Septembrio, or Thracias. It is only determined on two points—the one that the north should be the leading element, and the other that it had to blow its hardest.

It is quite a relief to turn into the quarry, as into harbour out of a rough sea, and to be free from that bitter, searching wind which takes away the breath when faced, and, when the back is turned, seems to force its way through all apparel as easily as if the thick overcoat were little more than chain armour. Here in the quarry, what a change is there! A few flowers still linger in this sheltered spot. The yellow ragwort is plentiful, and a few purple mallow flowers are visible among the green leaves. The soil, however, does not seem to be kindly for mallow, as the leaves, though numerous, are scarcely larger than penny-pieces; the plant crouches closely to the ground, and the flowers, instead of flaunting some three feet in the air, upborne by a stem like a walking-stick, are nestled among the leaves, and almost hidden by them. Richer colouring than the mallow is, however, there. The entrance to the quarry opens into 'Ragged Robin Lane,' and there, on the spot where the southern sunbeams can warm and the north wind cannot touch, is Ragged Robin himself, with just one or two rosy flowers yet unfaded. I suppose that the flower has held its own because, owing to its situation, nearly at the foot of a hill, the bottom of the quarry is always moist, and whatever warmth there may be it is sure to get. And in the middle of the quarry stands a solitary oat plant, tall, fair, and strong, its leaves broad and healthy, and its graceful pendulous spikelets waving gently in the slight breeze that can find its way into the quarry.

Thick and dark lie the fallen leaves, coloured with the yet unfaded reds and browns and yellows of autumn. Without moving I can note sycamore, maple, oak, Spanish chestnut, horse chestnut, beech, birch, elm, and ash. It is worthy of notice to remark how capricious are the trees in retaining or parting with their leafage, and how, when two trees of the same species stand near each other, one will be entirely bare, while the other will be half clad with fairly green leaves. This difference is evidently to be attributed to the particular soil into which the chief roots of the tree have penetrated.

The soil in this spot is exceedingly varied, all sorts of strata turning up close to each other. For example, the eastern and southern sides of this quarry are soft, friable sand, whereas the western side is rough conglomerate. Of the latter material, indeed, our hill is mostly composed. It is very healthy, no doubt, and has the advantage of creating scarcely any mud, so that, even after a long and steady rain, a lady can safely walk in the roads, provided that her boots be reasonably stout.

Still it has its disadvantages. The ground does well enough for trees and even shrubs, but it renders floriculture a heart-breaking business. Only the thinnest and poorest layer of soil lies on it, and even if abundant mould be added, the first heavy rain washes it all away, and a fine crop of loose stones comes to the surface. As for turf, it will not live on such a soil, but becomes covered with moss, and gradually dies off. After some ten years' experience, I have at last induced a lawn to exist; but then I had to dig away some eighteen inches of rubbish, put down a layer of good soil, than a thick layer of chalk, and then another of marl. Chalk absorbs water like a sponge, so that it retains the water which otherwise would have run to waste, and gives it out slowly to the roots of the grass in the dry weather.

Another disadvantage of such a soil is the abundance of pebbles, varying in size from a cherry to a plum, and nicely rounded for throwing. Consequently the boys, with whom this place abounds, and who, boy-like, are mostly at war with each other, and always with the rest of mankind, find themselves amply provided with weapons ready to hand. In the autumn, when the chestnuts are ripe, it is scarcely safe to turn a corner, or even to go near one, so perpetual is the fire that is kept up at the trees and between rival parties of boys.

The perpendicular sides of the quarry show this arrangement of strata very plainly. It is curious to see how the roots of the trees have restricted themselves to the shallow stratum of soil, so that they run almost horizontally. As the rain and wind beat away the upper portion of the quarry, the earth falls away from the roots, which hang down, waving loosely in the air; so when the strong wind attacks them they lash about like whips, and cut large semicircular grooves in the sand-wall against which they are blown.

Some trees seem to be little affected by this falling away of the soil. The elder, for example, retains its leaves bravely, and in one part has formed quite a rampart against the wind; so does the blackberry; while the elms are entirely stripped, the rooks' nests coming out black against the grey sky, whilst even the oaks have parted with their leaves, contrary to their usual custom of keeping them, though withered, until they are pushed off by the young foliage of the following spring.

In July last, among the many insects which thronged the quarry, I was greatly struck with the number of sand-boring and parasitic insects that buzzed about its eastern face, and so thought that such a day as this would afford a good opportunity for digging into the bank and seeing what the insects had done.

Even the face of the quarry has undergone a great change since July, not by the hand of man, but by natural means. The rains of many consecutive weeks have been dashed against it, run down it, and cut it into multitudinous meandering channels, while at the bottom of the quarry is a large heap of mud, composed of the soil and sand which have been washed down. Indeed, the view of the quarry showed admirably, on a small scale, how vast a work water does in changing the face of the earth. The strangest point about this channelled surface was the formation of numerous stalactites and stalagmites. A stalactite of sand seems rather a strange thing, but there they are and plenty of them. They are, of course, but small, only a few inches in length; but size goes for very little in Nature, and, when compared with the area of the whole globe, there is not very much difference between six inches and six feet. They fall to pieces at the slightest touch of the finger, and yet remain unhurt while the tempestuous wind is roaring above, and the air is full of heavy rain, whirling leaves, and bits of dry branches.

A portion of the eastern face has escaped rather better than the rest, and to that I directed my attention. It was literally covered with burrows, varying in size from eighteen inches to the eighth of an inch in diameter. The small burrows are evidently owing to the insects which were so plentiful in the summer. Chief among them were the Kentish bee (*Andrena pilipes*), a very local insect, hardly to be found in any other county of England except that from which it takes its name; the Sand-wasps (*Crabro* and *Odynerus*), and the lovely Ruby-tail flies (*Chrysis*), about all of whom we shall presently learn something.

The largest is that of a fox, and a very clever contrivance it is. After much search on the hill from which the quarry is cut, I found the other opening of the burrow. It is situated on the side of the hill, shaded by grass and bracken, and is so carefully concealed that, although I knew it must be situated within a limited area, I had some difficulty in finding it. Should the fox be run to earth, he would take refuge in this burrow, crawl by its means through the hill, slip down the face of the quarry, and be off to some other place of concealment. There are plenty of rabbit burrows; two of which are so close to each other as to bear a curious resemblance to the Thames Tunnel, especially as the rain has washed away the sand around them, so as to form a sort of arched recess, in which the two openings are seen side by side. Above them, and not far beneath the layer of soil, are a number of the sand-martin's burrows, now of course deserted, their inhabitants being in climates where they are certainly warmer, and, I hope, drier, than they would be here.

There are one or two mouse-holes; but these are of no consequence, and we proceed to those of the insects. First in size comes that of the Kentish bee. It is really a curious little insect. It bores horizontal tunnels some seven or eight inches in depth, each tunnel being about large enough to admit a common drawing-pencil. The insect itself would scarcely be recognised by those who had only seen specimens in a cabinet. Such specimens appear in their natural colours, *i.e.*, entirely black, while the bee, as it flies to its burrow, is entirely white. The fact is that in its state of grubdom the young bee feeds on the pollen of the thistle. The mother bee, after finishing her burrow, goes off to the fields, carrying away a quantity of the required pollen, and places it at the end of the burrow, together with the egg from which the future bee will emerge. The pollen being quite white, the bee is covered with it, just as a miller is covered with flour, so that she is quite metamorphosed for the time.

The sand being soft can easily be cut away with the knife, and, a grass-stem having been previously introduced into the burrow, there is no great difficulty in tracing it to the end. Sometimes, however, a large piece of sand breaks away and falls, carrying with it the whole of the burrow together with the grass-stem. At the end of the burrow may be found, at the proper time of year, the cocoon containing the bee-grub, and if it be carefully removed and placed in a box, the bee itself will make its appearance in due time. I have hatched out plenty of Kentish bees in this way. Although so local, it is a very common insect in this part of the country, where the soil is favourable. I am quite sure that, contrary to the habits of most insects, the Kentish bee has vastly increased in numbers since Kent was brought into the high state of cultivation which distinguishes the ' Garden of England.'

Before man brought his hand to bear upon the soil, the Kentish bee must have been sorely troubled to find a suitable place for its burrows. Sand very seldom forms itself into natural banks, and it is very rarely the case that a gulf is cut through the sand by the action of water, so as to leave a perpendicular bank on either side. Now, as the Kentish bee makes horizontal and not vertical burrows, it is evident that in the days when England was in the hands of savages, who made no roads and built no houses, the Kentish bee must have been much fewer in numbers. But, now-a-days, roads are cut through the sand-hills, and the sides of the cutting are filled with the bees' burrows. Sand, too, is urgently wanted both for building and agricultural purposes, and consequently almost every sand-hill has its quarry. It is most interesting in the bright summer time to watch these places, and see the white throng of Kentish bees flying into and out of their burrows, and making the air musical with their busy hum.

In the particular quarry of which I am writing, the Kentish bee has restricted itself to the upper portion of the sand, so that its tunnels cannot be reached without much difficulty. The lower part is occupied by the small burrows of the sand-wasp, which are placed so closely together that the face of the quarry looks very much as if it had sustained a series of volleys of No. 7 shot. Not exactly so, as we shall see, for in one place there actually is a group of shot-holes round the entrance of a rabbit-burrow, the gun having evidently been fired at the animal as it was making its escape. Shot-holes differ from those of the sand-wasps in this respect. The latter are quite circular, and their entrance is no larger than the diameter of the burrow itself, while the former are irregularly conical, the blow of the shot having always broken away a quantity of friable sand. Not a single shot remains in any of the many holes, the heavy leaden pellets having all rolled out of their conical beds.

To trace up the burrow of the sand-wasp is a difficult task. I find that the best plan is to select a spot about a foot square, in which the burrows are very numerous, and then to pare away the sand in thin slices. If this be done neatly and carefully, the whole of the burrow can be laid down open from mouth to end. Mostly they run horizontally, like those of the Kentish bee, being driven at right angles to the face of the sand-bank, but some of them make a sudden curve, when they have gone a few inches into the sand, run for a little distance parallel with the quarry face, and then resume their former direction.

Suddenly we come upon a small lump of something black and fluffy, looking much as if a small pinch of black cloth teasings had been rolled into a little cylinder and pushed to the bottom of the tunnel. We carefully get it out with the point of a penknife, and slip it into a box, so as to prevent it from being blown away by the wind. Presently another and another of the black lumps is discovered and transferred to the box. Presently we come to another lump, which is pale brown instead of black, and place it with the others. Now, having preserved as many specimens as are

wanted, we make our way homeward through the ɪain
and wind, and proceed to the microscope, in order to
ascertain the precise character of the fluffy lumps taken
from the burrows.

The day is much too dull and dismal to afford
sufficient illumination, so the lamp is lighted, and one
of the black objects placed under the half-inch glass.
The first glance detects its nature. It is composed
entirely of fragments of little flies. Black, shining
bodies, heads, and severed wings are clustered thickly
together, the wings shining out in every colour of the
rainbow, amid the *débris* with which they are surrounded. The sand-grains look like lumps of sugar-
candy, the withered, red-brown eyes still show their
thousands of hexagonal lenses, the black, hairy legs and
fragments of bodies lie about in utter confusion, while
the wings, though broken from the body and mixed
with sand and all kinds of miscellaneous rubbish, flash
and glitter in ripples of crimson, green, gold, and
azure. Gauzy and delicate as they are, they have survived the body to which they were once attached, and
have not lost one whit of their former beauty. One
fly presents a very curious aspect. It is a little black,
round-headed fly, quite shrivelled up and withered. It
has lost all its legs, but it retains its wings, and adheres
to the general mass by the very tips of those organs,
projecting itself forward, and looking like a tiny black
imp sustained on bright, glittering, many-coloured
wings that would do credit to a fairy. Altogether,

one of these insect masses reminds me much of the 'pellets' which are found so abundantly in owls' nests, and which are composed of the skin, bones, and teeth of mice, and the hard limbs and wing-cases of beetles.

The black lumps are all composed of the same materials, so we pass to one of the brown masses. No opalescent patches of colour betray the presence of wings, but projecting from it on every side are long, crooked legs, covered with sharp, brown, curved spikes, showing in a moment that they are the legs of spiders. All these brown masses are alike; the spiders are apparently of the same species, and all nearly the same size. After examining a considerable number of specimens, I can only find two materials for these masses, namely, spiders and flies, and in no instance is there a spider among the flies, or a fly among the spiders. Now why were these creatures buried in the bottom of these tunnels, and why are they so shrivelled and dismembered? They were placed there by the sand-wasp as food for her future young, just as the Kentish bee stores her burrow with pollen. Sand-wasps in all their stages of existence are carnivorous, and so it is necessary to supply the young with the appropriate animal food.

There are very many species of sand-wasps, and each chooses some particular insect as food for its young. Many prefer flies, some furnish their young with aphides, and others choose beetles. Even the little hard-bodied turnip-beetles (turnip-fleas, as they are often called, on account of their small size and powers of jumping) are used for this purpose. How the little sand-wasp grub manages to eat them is more than I know, but perhaps the hard integuments may be softened by the damp of the burrow. This, however, is merely conjecture.

There is yet one insect to be accounted for. I have already mentioned the ruby-tail flies that in July were flitting so anxiously over the face of the quarry, their burnished crimson and blue mail flashing in the sunbeams like living jewellery. They were on a somewhat similar errand to that of the bees and wasps, but they carry it out in a different manner. They are parasites on the sand-wasps, and just as the sand-wasp grub eats the flies, so the larva of the ruby-tail eats both the sand-wasp grub and all its store of food. From observations that have been made on the habits of these insects, the larva seems at first to suck, rather than to eat, the unfortunate grub on which it feeds; but, having extracted nearly all the juices, proceeds to devour the other portions of the body.

The mother ruby-tail is wonderfully persevering in her attempts to insert an egg into some other insect's nest. Sometimes the rightful owner detects the intruder, and then the latter generally suffers for her deeds. When attacked by her angered foe, she usually tries to shield herself by rolling her body into a ball and lying motionless. Even this ruse, however, does not always save her, and she loses her life, together with her hope of providing for a future generation.

Considering the size of the ruby-tail, it can contract itself in a really wonderful manner. Some little time ago, on a bright day in early spring, I was looking at some rough palings upon a park fence, and was examining the little holes made by the Scolytus and similar beetles. The palings happened to face due south, and as the meridian sun shone on them, a ray penetrated into one of the holes and I discovered something blue within. I proceeded to cut it out very carefully, and there found a ruby-tail completely doubled up, like a hedgehog, within a hole scarcely large enough to admit a No. 5 shot. In the same row of palings I found plenty more specimens, all alive, and very much perplexed at being so unceremoniously ejected from the resting-place in which they had passed the winter.

UNDER THE BARK.

MARCH.

THERE is a time for all things, and this is the time for that pursuit so dear to the heart of all entomologists, hunting 'under the bark.' And well may it be dear to him, for, putting aside the fact that 'under the bark,' and there only, are found some of the rarest insects that can enrich a cabinet, the pursuit is in itself one of singular fascination. By withdrawing the curtain of the bark we are admitted, as it were, on the stage whereon Nature acts her ever-varying drama, and indeed penetrate behind the scenes of the theatre. We trace insects throughout the various stages of their existence, and see how, by regular degrees, the fat, white, round-bodied, slow-moving grub is transformed into the active, ample-winged, long-legged beetle. Then there is all the fascination of the lottery attendant upon the search under the bark, and every fresh tree or stump contains within it new elements of amusement. That there will be something under the bark is absolutely certain; but what it may be no one can tell. There may be, perhaps, nothing but a common woodlouse or centipede; but there may be, and very likely will be, some insect which the searcher never expected to find, and for which he has been long looking. Without further preface I will give an account of a few hours just spent in looking under the bark.

Close to my house—only across the road, indeed—there is a piece of ground which at one time was thickly planted with oak, birch, and fir trees, but which has been of late years partially cleared; the stumps of the trees being in most cases left standing, so as to project a foot or so above the ground. These trees are on the upper part of the ground, while on the lower, through which runs a tiny but permanent brooklet, some willow trees are planted, one or two of them being very old. In this ground I lately spent between two and three hours, armed with a mortice chisel, a pair of forceps, and a laurel-bottle, *i.e.*, a bottle in which are some young crushed laurel leaves, the odour of which is fatal to beetles, and prevents them from eating each other. At this time of year, however, no young laurel leaves are sufficiently grown, so I had to make a substitute for them by putting a little Easter's insect powder at the bottom of the bottle, and covering it with a finely perforated card. This plan answered so well that I shall try it through the season, for not only was the odour of the powder fatal to the insects, but it did not stiffen their limbs, which is the result of laurel-leaf vapour.

As I have already mentioned, this is a good time for searching under the bark. The severe frosts of winter have passed away, so that the fingers of the searcher are not chilled into uselessness, a circumstance which is very apt to occur when the enthusiastic entomologist pursues his task in mid-winter. Moreover, towards the spring a vast number of tree-inhabiting insects become developed, and make their way towards the open air before they undergo their last change, so that at this time of year we may find almost immediately under the bark many insects which at other times would be buried deeply in the wood.

Taken roughly, all the creatures which are found under the bark may be divided into two classes, namely, those which have resorted there for shelter during the cold months of winter, and those which feed upon the bark or the substance of the tree itself. The former can always be found under the bark of old trees, especially oaks and willows. The latter, however, are the most prolific in insect life. In many old willows the bark is slightly separated from the trunk for many feet, and although no external sign be given of this fact, the hollow sound which is returned when the outside of the tree is tapped is a sufficient proof. On carefully removing one of these sheets of bark from the tree a most extraordinary sight is often presented.

The space between the bark and wood is a vast camp of insect armies, their white and glittering tents being set so closely together that there is not room for a finger's tip between them. Under the bark also flourish certain colonies of flat white cryptogams, which spread themselves in fan-like rays, and almost rival the silken insect tents in whiteness. Now and then comes a circular tent, through which can be seen a quantity of little yellow globular objects. The character of the silk tells us that the nest is certainly that of a spider, and we just pull off a little of the cover to get a better view of the eggs. Scarcely has the tip of the forceps stretched the silken roof than a simultaneous stir becomes apparent through the eggs, and all at once they suddenly start into life, unpacking in some mysterious way the limbs which had been folded round their globular bodies, and all running about as busily and aimlessly as the inhabitants of a disturbed ants' nest. In fact, the seeming eggs are not eggs at all, but very young spiders which have only just been hatched and are waiting for warmer weather before they make their appearance in the world.

This same space between the bark and the wood is a favourite resort of many moth-caterpillars. Led by instinct they proceed to the tree and climb up the bark, seeking for some recess in which to pass their short period of helpless existence. In comparatively young trees they content themselves with the crevices formed by the rugged and knotty bark, but in old trees, such as have been described, they manage to discover some aperture through which they crawl into the large sheltered space and there spin their silken home. Careful investigation shows that, however safe such a retreat may be for the insect while in its pupa or chrysalis condition, it is little more than a trap for the perfect insect. For not only are spiders' eggs to be found 'under the bark,' but spiders themselves also take up their residence there, and find ample subsistence in the many insects that have found their way under the bark and cannot find their way out again.

Only two or three days ago I found, under the bark of an old willow tree, the remains of a beetle (*Pristonychus terricola*) which had fallen a victim to a spider. Unfortunately the edge of the chisel came upon it and damaged the specimen, or I should have cut it out and preserved it for my museum, as I never saw anything more curious. In the first place, that the insect had been caught by a spider was evident from the fact that it was bound to the tree by spider web. In the next, it was laid on its back with the limbs, jaws, wings, and wing-cases separated and displayed with as much regularity, in spread-eagle fashion, as if it had been prepared by an entomologist as a specimen of insect anatomy.

Now, that any beetle should have been so treated is remarkable enough, but it is still more wonderful when we remember that the insect in question is one of the predacious beetles and measures three-quarters of an inch in length, so that it appears to be much more likely to eat the spider than to allow itself to be eaten.

Of all the insects which hibernate in the crevices of the bark, by far the greater number seem to be the chrysalides of various moths, which, as a rule, hide themselves so well that they need a practised eye to see them, and even though the greatest care be taken are often accidentally destroyed. It is extremely provoking, after selecting an apparently safe spot for the chisel, to see a white creamy fluid run along the blade, and then to know that the tool has passed through the body of a chrysalis which has hidden itself so cleverly as to escape observation.

The most successful of these hiders is the Puss Moth (*Cerura vinula*), the chrysalis of which lies hidden in a singularly ingenious cocoon. When the caterpillar is full fed it crawls to the trunk of the tree and looks about for a crevice in the rough bark. Into this crevice it insinuates itself, and begins at once to nibble the bark into tiny chips, which it fastens together with the silk-fluid discharged from its spinnerets, and so makes a cocoon which completely shelters it. Owing to the materials of which the cocoon is made, it exactly resembles the bark and can scarcely be distinguished from it, and as the caterpillar took care to retire into the crevice before spinning, the surface of the cocoon does not project beyond that of the bark in general. Very often when the eye fails in detecting a cocoon the touch succeeds, the material of the cocoon being soft; but this is not the case with the Puss Moth, whose cocoon is much harder than the bark of which it was made, the silk-fluid forming a wonderfully firm and tough cement.

As for woodlice, millipedes, armadillos, and centipedes, they swarm under the bark, especially the wood-lice, whose dried and whitened skeletons can be seen by hundreds, showing at once their crustaceous descent. Earwigs also are sure to appear in great force, and, as is their wont, do not lose their presence of mind when disturbed, but make their way instinctively for the nearest crevice, and wriggle their lithe bodies out of reach almost before they have been seen.

Under the bark are also the relics of other creatures. For example, in one willow tree, the nuthatch and the squirrel have both left their marks in the shape of sundry hazel nuts. There is no difficulty in distinguishing the work of these two creatures. The nuthatch wedges the nut firmly into a crevice of the bark, and hammers rapidly and perseveringly at the point until the nut is split in two as neatly as a boy could do it with his knife. The bird then goes off with the kernel and leaves the halves of the shell where they happen to lie, some of them being still fixed in the bark, some lying on the ground, and some having slipped between the bark and the wood. In this place the nuthatch abounds, so that there is every opportunity for watching its habits. The squirrel treats its nuts in a different manner, first gnawing off the tip and then splitting the shell with its chisel-shaped teeth. Sometimes it gets hold of a bad nut, and after nibbling at the tip throws it away. In the same tree that has just been mentioned were several bad nuts, each of which had been tested by a squirrel and then thrown into the hollow between the tree and the bark.

There are plenty of beetles which find a shelter under the bark during the months of winter. While taking off one of the bark strips I saw something black wriggling violently under the chisel, and presently saw that it was a fine 'Devil's Coach-horse' (*Goërius olens*), which had evidently attacked the chisel after the manner of its fearless kind, and got itself caught between the blade and the wood. This is not a pretty creature, but it is a wonderfully courageous one, and will fight any antagonist without the least regard to size. These beetles are very common at Margate, living in the clefts of the chalk cliffs. I found one of them at the foot of a flight of stone steps, and was exceedingly amused at the manner in which it attacked my stick. It retreated fighting to the very top of the stairs, keeping its front well to the enemy, and acting on the offensive as well as the defensive whenever it found a chance. These are useful beetles to the gardener, as they feed upon many injurious insects, and I rather fancy that the unfortunate individual which was caught by the chisel had found its way under the bark as much for the sake of food as of shelter.

Having thus examined the willow trees, I ascended the hill and turned the chisel upon the stumps of the fir trees. These were, as I thought they would be, very prolific in insect life. In all cases I began by gently removing the outer bark, and in the first stump that I opened a gleam of rich metallic purple caught my eye. It was well below the bark, and buried in the soft decayed wood, and had it not been that a bright ray of sunshine happened to light upon it, I should perhaps have missed it altogether. On picking away the wood a fine specimen of the Purple Ground Beetle (*Carabus catenulatus*) was disclosed—in my opinion one of the handsomest of our British beetles, with its rich purple thorax, and the purple edging of its beautifully sculptured and elegantly shaped elytra.

To find one of these beetles is easy enough, because it is one of our commonest species. But I certainly never expected to find it in such a position. It is one of the predacious beetles, and both in its larval and perfect condition is a destroyer of other insects, so that unless it fed while in the larval state upon the wood-eating insects that inhabited the stump, I can scarcely account for its presence. It was not a beetle which had merely hidden itself under the bark by way of finding shelter, for the beautifully perfect condition of the insect showed that it had not as yet undergone any battle with the world. Moreover, it was hidden rather deeply in the wood, and was not merely lying under the bark. I think, therefore, that it must have fed while in the larval state upon the insects which inhabited the stump, and have crawled into the spot where it lay for the purpose of undergoing its transformation. Under the bark of the same stump, and in many others, were found various little beetles, which are popularly known as Sun-beetles or Sunshiners.

Carefully opening another stump, and removing half an inch or so of the rotten and damp wood, a slight movement caught my eye, and in a short time an antenna, evidently of a beetle, was seen gradually working its way to the light. Presently another antenna appeared, and then the head, which at once proclaimed itself as that of one of the wood-burrowing, long-horn beetles, called *Rhagium bifasciatum.* I do not know that it has any popular name, as is indeed the case with most beetles, however common they may be. This is a pretty, though soberly coloured insect, long bodied, long horned, long legged, and having a bold and sharp spike on each side of the thorax. To the unassisted eye it is only blackish gray, with four diagonal, cream-coloured marks on the elytra. But when a powerful light is concentrated upon it, and the magnifying lens is employed, the colouring assumes a very curious aspect. The elytra seem to be made of black glass, ribbed, and covered on the surface with a multitude of tiny white specks, while the cream-coloured marks appear to lie quite beneath the surface, as if they were painted under the glass.

I was very glad to find this beetle, having tried in vain to discover one of its curious nests, but, though three insects were in the same little stump, I could not find a perfect nest. At last, however, in another stump

I succeeded in finding the nest, cut off the stump with a saw, and brought it home.

Just before these beetles are about to change into the perfect state they make for themselves an oval cell, so shaped that the head of the insect is upwards. This cell is lined with strips of wood, which are torn away by the jaws of the larva and arranged regularly, like the tiles of a house. The nest that is now before me is a little more than an inch in depth, and on the outside is an inch and a quarter in length by three-eighths of an inch in width. This diminishes, however, both in length and breadth in proportion to the depth, so that at the bottom, where the insect reposed, it is only five-eighths of an inch in length and a quarter of an inch in width, just large enough indeed to hold the beetle with its limbs and antennæ packed tightly to the body. The insect which made this nest differed from all the others in one respect. In the other cases the beetles seemed only too anxious to escape from their dark home and pass into the open air, while this one persisted in adhering to its nest, and, as the light was admitted, seemed to prefer darkness, pressing itself into the farthest recesses of its cell.

Another stump disclosed a really wonderful scene of insect life. On stripping off the bark of a small stump, barely eight inches in diameter and about as much in height from the surface of the ground, a large colony of the Yellow Ant (*Formica flava*) was suddenly exposed to the light. The insects had the strongest objection to the inroad upon their premises, and ran about actively in all directions. Their habitation was elaborately made of small particles of earth, which had been built together after the fashion of ants, and had been arranged between the bark and the wood so as to form a perfect labyrinth of soil cells and passages. I was really sorry to have broken into so elaborate a piece of insect architecture, but the mischief had been done, and was aggravated by a brisk and decidedly cold wind which had just sprung up, and which blew the unfortunate ants about in a way of which they did not at all approve.

This species of ant is very common, especially on heaths and similar places, and has the power of varying the structure of its nest so as to suit all conditions. On open ground it builds little hillocks, which, fragile as they appear, are quite capable of throwing off the rain. If, however, it can find a flat stone, it takes advantage of so good a shelter, and makes its habitation immediately beneath it, while in the present instance it had run up its chambers from the earth and extended them between the wood and bark of the stump. The bark was very close to the wood, and the insects had gained the requisite space by making shallow cavities in the decaying wood. During a severe winter these ants carry their habitations deeply into the ground, and make chambers, sunk well beneath the surface, communicating with each other by passages some four inches in length. It is in these nests, by the way, that some of our rarest beetles are discovered. The colour of the insect is bright but pale yellow, the larger workers being brighter in hue than the smaller.

Having already done as much mischief as could be done, I had no scruple in removing the remainder of the bark. To my astonishment, another ants' nest was disclosed, but that of a different species, namely, the Jet Ant (*Formica fuliginosa*). Thus we have the curious fact that on opposite sides of the same little stump were two flourishing colonies of two different species of ant, neither interfering with the other, and both so completely concealed that no traces of them were seen until the bark was removed.

When their house was thus broken open, the ants showed at once the difference in disposition as well as in form. The yellow ants ran about in a state of great perturbation, and although they could do but little appeared to do a great deal. They were very angry too, and one of them, when put into the bottle, attacked a sun-beetle, grasped one of its antennæ with a hold like that of a bull-dog, and so died under the influence of the poisoned vapour. As a memorial of the occasion, I intend to place in my cabinet the beetle with the dead ant still griping the antenna between its jaws.

The jet ants displayed no such fussiness, but took matters very coolly indeed. At first they seemed to be surprised into something like activity, but they soon appeared to make up their minds that there was no use in troubling themselves more than necessary. So they quietly slipped away under cover, some dropping at once to the ground and so escaping into the recesses of the nest, and others crawling very leisurely down the ruins of their home and disappearing into the deep-lying cells. Still, quiet, and almost sluggish as were their movements, they answered their purpose wonderfully well, for the retreat was made so quickly that almost before I had recovered from my surprise at finding this second nest so close to the first scarcely an ant was to be seen.

The jet ants carry this easy-going style into ordinary life, and may be seen in the summer time doing what few ants do, namely, idling. They have a way of assembling together in considerable numbers just outside the nest and remaining quite still in the sunshine, instead of running about and working as do most other ants, whether the sun shine upon them or not. I need scarcely say that both these ants are well worthy of examination under the microscope. All insects are worthy of such an examination, but some seem to be more worthy than others, and of such are these two species of ants, so different from each other in form, colour, and disposition.

The same fir-wood is also much frequented by the magnificent insect called *Sirex gigas*—the insect that is so common in the neighbourhood of fir woods, and is so often mistaken for a hornet. When the bark is removed from the tree, the holes made by the sirex in its larval state are very evident; but as the insect is an inhabitant of the wood itself, and is seldom to be found under the bark, I pass but lightly over it in the present instance. Under the bark is a favourite resort of many of the weevil tribe, and those which do worst harm to fruit-trees are mostly in the habit of hiding themselves during the winter in the crevices of the bark. So all growers of fruit-trees will do well during the winter to search carefully under the bark of the older trees, and to fill up the crevices with some greasy composition, which will smother the beetles as they lie in their hiding-places. Stripping off the bark is much recommended, but I doubt its efficacy, inasmuch as these beetles always let themselves fall to the ground when they are alarmed, so that the greater number would escape when they were disturbed. The greasy composition, which can be laid on with a brush, does no harm to the tree, and very effectually smothers the insects that are lying hidden ' under the bark.'

MRS. COATES'S BATH.

Mrs. Coates's Bath is, I am happy to say, a bath no longer, but subserves a better purpose. What it was in the old days, and what it is now, I proceed to explain.

One of the most delightful privileges of a practical naturalist is to possess extensive grounds of varied character. The next best thing is to live close to such grounds possessed by a friend who allows free range over them. In the same grounds that were mentioned in the paper entitled 'Under the Bark' is a small pond situated at the bottom of a rather steep dell close to the house. Somewhere about the beginning of the present century a certain Mrs. Coates inhabited the house, and very judiciously converted a piece of swampy ground into a convenient bathing-place, by having a pond dug and paved, and the spring which saturated the ground led into it at one end, and out of it at the other, and so conveyed into a brook which just skirts the grounds. For many years, however, the place has been disused as a bath, and merely serves as a pretty object to the eye. A week or two ago the idea struck me that the pond was likely to be rich in animal life, inasmuch as it is completely sheltered from the north-east wind, which is so hated by insects, and only open towards the south, allowing the meridian rays of the sun to fall daily upon it. So I fished in it for an hour or two, with a little net, and found that my impression was true. It is absolutely impossible, in so limited a space, to describe, or even to mention, all the creatures which I found in that tiny pond, but the following are some of the most characteristic inhabitants of ' Mrs. Coates's Bath.'

There were plenty of newts. Now, these are really very pretty creatures, especially in the breeding season, when the males put on their nuptial splendours. Like many birds, they only assume their best dress for a short period, and when that brief period is over, they can scarcely be distinguished from their more sombre mates. The chief and most conspicuous portion of the nuptial dress of the male newt is a sort of fin which runs along the back, and looks something like a cock's comb. It is deeply notched and toothed at the upper edge, and, as it is extremely delicate, it waves about in the water in graceful accordance with the movements of the animal. My little boy took some of these newts home, and, in the innocence of his heart, showed them to the gardener. The man was horribly frightened. He jumped back and absolutely yelled with terror. He, keeping at a safe distance from the dread beasts, told the boy that ' the effet was the most pizenous thing as is,' and that he had known lots of people lie down in the grass at haymaking time, when they were bitten by effets, and then they swelled up and went on swelling till they died. Fortunately, the boy was too well taught to believe the man, and his terrors and warnings only afforded the keenest amusement.

The newt is an interesting animal to keep. In the first place, it is very graceful as it swims, and its pretty colours and brilliant eyes show out much better in the water than on land. Then it has a very curious manner of depositing its eggs, doubling them up in the leaf of some plant, sometimes, though not always, a plant which is growing in the water. There is now before me a blade of grass which I found in the pond. It is neatly doubled in two, and in the fold is one of the little translucent eggs of the newt. When these eggs are hatched, little tadpoles issue from them, almost exactly resembling those of the frog, having similarly large heads, and long tapering bodies.

They do not show their individuality until their legs begin to appear, when the distinction is at once evident. In the frog-tadpole the hind legs are the first to appear, but the reverse is the case with the newt. As they increase in size the distinction becomes more apparent, for the tail of the frog-tadpole is gradually absorbed into the body, while that of the newt increases in length. The newt, in fact, differs but little in structure from the frog, except that it retains its tail throughout its life. I find that in captivity the newt changes its skin oftener than it would do if left at liberty, and that if the water in the vessel in which it is left be changed, the newt generally casts its skin within a short time.   This envelope is drawn from the body in an almost perfect state.   It is exceedingly fine, like goldbeater's skin, and, if a card be slipped beneath it as it floats in the water, the skin can be spread out on it, and so removed, dried, and preserved as a specimen. Two such skins are now before me, and very pretty objects they are, every toe of each foot being quite perfect, and looking like fairy gloves.

As to aquatic insects, the water swarms with them, and the number of water-beetles alone that I found there is prodigious.   This not being an entomological work I do not intend to give a list of the insects found in this little pond, but will only mention some of the principal species.

There was the great water-beetle (*Dyticus marginalis*) in plenty.   This, in common with all of its kind, is not an eligible inhabitant of an aquarium.   If two or three be placed in a vessel with other inhabitants of the water, they immediately begin eating their fellow prisoners, and, having finished all the smaller creatures, attack each other, the strongest killing and eating the weakest.   I have before me a very fine male Dyticus in a pickle-bottle, where I was compelled to banish him in consequence of his voracity.   I feed him mostly on blue-bottles, which he seizes between his powerful fore-legs, and devours in a very short time.   At first he was rather puzzled with the flies, they not being his usual prey, but he now knows how to manage them, and a fly scarcely touches the surface of the water when it is seized and borne below, held firmly in the jaws of its captor.

It is an insect full of wonders, and contains in itself the elements of more than one mechanical invention. In the first place it is a living diving-bell. Like all insects it breathes atmospheric air by means of tubes, which permeate the whole of the body. The apertures by which these tubes communicate with the air lie on the upper part of each side, under the wing-cases. Now these wing-cases, or elytra, are convex, while the upper part of the body is flat, so that there is a space between the wing-cases and the body. Every now and then the beetle comes to the surface of the water, protrudes the end of the body, draws in a supply of air into the space between the elytra and the body, and dives again, the elytra fitting so closely to each other and to the sides that the air cannot escape. Sometimes, if the beetle be not alarmed, it will remain at the surface, with its head downwards, and its body balanced by its extended swimming-legs, and on a calm day quite a number of water-beetles may be seen thus suspended.

The swimming-legs which have just been mentioned are themselves very wonderful examples of structure. They are so made that the only movements which they can perform are those of swimming, and they are fringed with stiff hairs so set that when the leg is struck against the water the hairs stand out and act like the blade of an oar ; while, when the limb is bent back for the next stroke, they are drawn through the water from their roots to their points, and so offer the least possible resistance.

The first pair of legs of a male Dyticus are worthy the closest possible examination. On the feet of each of them is a round disc, which, when magnified, is seen to be made of three joints, flattened and dilated. Their under surface is covered with a vast array of suckers, one of which is very large, two of moderate size, and all the rest very small, and set on foot-stalks. With these suckers they can hold so tightly that they can crawl up a pane of glass by their aid, and hold so firmly that a rather sharp pull is required before they can be detached. A few days ago I noticed that one of these water-beetles had remained at the bottom of the vessel for a long time, and, on closer examination, found that it was dead. I took hold of it to remove it, when, to my astonishment, the body came away in my hand, the two fore-legs still clinging to the glass. The beetle had evidently been dead for some time, and was semi-putrid, but yet the suckers held on as firmly as during life.

Owing to the great length and peculiar jointing of the swimming-legs, the beetle is a bad walker, though it is a good flyer and a better swimmer. If placed upon the ground it crawls awkwardly about, and seems to have little power of directing its course. Should it fall on its back on a smooth surface it gives a series of wild kicks with its long hind legs, the action being precisely the same as in swimming, and both legs being used simultaneously. If the surface be perfectly smooth, such as a plate or a piece of glass, the insect only spins round and round, and after a short time seems to be seized with despair, and lies perfectly motionless.

Though the beetle can do no harm, and may be taken in the hand without fear, I do not recommend indiscriminate handling, and this for two reasons. In the first place it is wonderfully strong, and has a way of forcing itself backwards through the hands, so that a double-headed spike at the base of the swimming-legs is apt to prick the fingers rather smartly. In the next place when held it ejects a whitish fluid, which issues from the junctions of the head, the thorax, and the abdomen, and which has a strong and very un-pleasant odour.

The larva or grub of this beetle is quite as formid-able and ferocious as the perfect insect—which it does not in the least resemble. It is long-bodied, the body swelling out in the middle, and tapering gradually to the tail, at the end of which are a couple of diverging fringed leaflets, which are attached to the respiratory organs. The head is large and broad, and armed with a pair of exceedingly long and sharp jaws, curved like a reaper's sickle, and having a very sickle-like aspect. The legs are long and slender, and the colour is a pale brown. It moves by two modes of progression. It can use its legs for walking, and does so when it is trying to crawl leisurely, but when it desires to move with any swiftness it causes its body to undulate like the movement of an eel or a serpent, and so gets along at a good pace, its legs being used merely as balancers to its body.

Its character is best seen when it is at rest. It bends its body nearly at right angles, and ascends to the surface of the water, upon which the fringed leaflets of the tail are spread so as to enable the creature to breathe at ease. It thus hangs, as it were, suspended by these leaflets, with its head downwards, its monstrous jaws wide open, and its long legs spread, so that it forms a perfect living trap, ready to close on any unfortunate creature that may come in its way. Being rather a wary creature it escapes the net unless proper precautions be taken. I have always found that the best plan is to stir up the mud of the pond, and then to sweep the net rapidly through the turbid water, thus catching the Dyticus larva before it sees its danger. In 'Mrs. Coates's Bath' are many of this beetle's kinsfolk, but the manners and customs of all are so similar that one will suffice as an example of them all.

Perhaps the reader may think that there is not much to be seen in the common whirlwig, or whirligig beetle (*Gyrinus*), which may be seen in vast numbers on the surface of the water, performing its mazy dance in any sheltered spot. Summer or winter seem to be the same to the whirlwig, and even in the cold days of winter a gleam of sunshine will bring out the whirlwig beetles in any spot wherever the ice is not formed, and they will dart about as merrily as if the July sun were pouring its hot beams on them. I need not say that there are plenty of these beetles, because there is scarcely a piece of water larger than a puddle in which they may not be found. A depression in the ground which has been dry for months, and suddenly filled with water by a rain-storm, will have whirlwigs in it before many hours have passed. The fact is that these beetles, like those which I have just described, have large and powerful wings, and can use them with great ease. They can take a flight from the surface of the water—a fact which I believe has not hitherto been noticed, or at all events not published. I found it out only a few weeks ago.

While stooping over the water, and admiring the rapid movements of the whirlwig beetles, one of them suddenly darted up, struck me on the nose, and fell back again into the water. If the beetle were half as much astonished as I was, it must have been very much surprised indeed. Wishing to see how this feat was achieved, I took a number of the beetles, and put them into an aquarium, thinking—and, the result proved, rightly—that they would soon be tired of their limited space and would take to wing. After whirling about for a little time, some of them crawled up the glass sides of the aquarium, while others darted into the air and took to flight. They did it by striking the water violently with both swimming-feet at the same moment, and thus jerking themselves several inches into the air. Almost simultaneously with the spring, they spread their wide wings, and flew off with incredible speed.

There is an old fairy tale about three sisters, who had respectively one, two, and three eyes, the elder and the youngest treating their sister very contemptuously because she had two eyes like people in general. Now, a whirlwig beetle goes one step beyond them for it has four eyes, two above and two below—two to see below the water and two to see above it. Of course the beetle has in reality a vast number of eyes, like most insects, but those eyes are divided into four masses instead of two. The reason is this. The insect is continually scurrying about on the surface of the water, watching for prey, and if its eyes were constructed in the ordinary fashion it would only be able to see either above or below the surface, according as its eyes happened to be placed. In order to be able to see distinctly any object below the surface of the water, its eyes must be submerged; and in the eyes of this little beetle we find the principle of that well-known instrument, the water telescope. This is used for the purpose of looking into the water, and is simply a tube with a plain glass fixed water-tight into one end. When the glass is pushed under the surface of the water, and the eye applied to the upper part of the instrument, objects can be seen with great plainness, the vision not being obstructed by any ripples on the surface of the water.

In the larval state this is a very peculiar creature. It is long-bodied, with a blackish head, and along the sides of the body are delicate white filaments, which are the 'branchiæ,' or gills, by which the creature breathes. Respiration is effected by a continual passage of water over the gills, and in still water this object is achieved by a very constant undulation of the body, so that the gills necessarily are brought in contact with fresh particles of water. By means of the same undulations, the larva urges itself through the water, just as has been mentioned of the Dyticus larva, and so, by the mere act of progression, increased power of respiration is obtained. In perfectly still water, the creature is never quiet for a moment, but keeps up a perpetual undulation of the body, during which the little gills have a most graceful appearance, as they float like silver threads on either side of the body. Sometimes the larva obtains its supply of oxygen by ascending a few inches by forcible undulations, and then allowing itself to sink slowly to the bottom, the delicate branchiæ being spread out on either side, and acting as floats to prevent it from sinking too fast.

In 'Mrs. Coates's Bath' are numberless water-boatmen (*Notonecta*) of various species and in all stages of existence. We will, however, content ourselves with the commonest and largest species. The insect derives the popular name of water-boatman from the fact that it lies on its back, the sharp edge of which makes a very good imitation of a boat-keel, and rows itself by its long swimming-legs, which are nearly straight, and, with their bristle-fringed ends, look exceedingly like oars. In fact, we have no oars that can in any respect approach in efficiency the swimming-legs of the water-boatman, with their invariably correct action, and their self-feathering blades. The name of Notonecta or back-swimmer is given to the insect in consequence of the habit of turning on its back when it swims.

These insects are not beetles, though they are often thought to be so. They belong to another order of insects altogether, and will give very tangible proofs of this fact if carelessly handled. Anyone who has caught one of the predacious beetles may expect a sharp nip with the jaws if he does not take care of himself. But the water-boatman, in common with the rest of its kin, is furnished with a sharp and strong proboscis, which it will drive deeply into the fingers of its captor if it gets a chance. Like the whirlwig, the water-boatman is able to take flight directly from the surface of the water, and does so in a very similar manner, leaping out of the water by a violent stroke of its swimming-legs, and then spreading its wings before it falls back again. When on the wing, it flies with a deep humming sound, very like that which is produced by the humble-bee.

Its respiration is carried on much in the same way as that of the water-beetle already described. On a calm day, if 'Mrs. Coates's Bath' be approached cautiously, so that a heavy step does not communicate itself through the land to the water, and that no shadow be thrown upon the insects, whole fleets of water-boatmen of all sizes may be seen floating with their heads downwards, the swimming-legs spread wide by way of balancers, and the tips of their bodies just protruding from the surface. A hasty step, however, a sudden movement, or a shadow, even of a passing bird, thrown on the water, will alarm the insects, and they will scurry off in all directions.

By watching these insects very carefully in a bottle, and keeping that bottle constantly before my eyes on my desk, I have been enabled to observe the course which the air takes in respiration, the partly translucent wing-cases enabling the bubbles to be traced as they pass like globules of quicksilver under the wing-cases and finally into the water. The air is taken in at the end of the tail, and introduced into the space between the wing-covers and the body. It is then gradually drawn forward until it reaches the base of the wing-covers, and is lastly forced out just where the wing-covers fit against the breast. When the insect is perfectly quiet the process may be seen going on with perfect regularity, the air being taken in near the tail, working its way under the wing-covers, and at last squeezed out near the breast, when it ascends in bubbles to the surface. In this position the water-boatmen are accustomed to wash themselves. They are as cleanly as cats, and perform the operation of washing in a very similar manner, leaving not a limb nor a part of the body untouched. Sometimes they will rest *on* the surface of the water, but this time with their backs upwards, their wing-cases half opened, and their wings partly unfurled. I never saw them assume this attitude except when the sun was shining directly on them, but I have in that case seen thirty or forty at a time sunning themselves in this curious attitude, which has all the effect of a disguise, and makes them look quite different insects.

I am sorry to say that water-boatmen are very predatory characters, and that they have a great fancy for preying upon the water-gnats, as they are called, those slight, dark-coloured, long-legged insects that run about on the surface of the water as if they were on land. They seize on the unfortunate insect, clasp it tightly to them with their fore-legs, drive their beaks deeply into its body, and suck out all its juices, afterwards rejecting the body, which to the eye seems to have undergone no change at all, and only to have been killed by the wound. The water-boatman takes from five minutes to a quarter of an hour to suck a single water-gnat, and carries it about almost pertinaciously, not even loosening its hold if alarmed and forced to dive.

Few facts have struck me more forcibly than the peculiar life which is led by this and other aquatic creatures. As a rule they are essentially predacious. Taking merely those which have been mentioned, we have the newt, which eats all kinds of water inhabitants, provided they are not too large, and is in its turn often subject to a fierce attack by the Great Dyticus, and has a bite or two taken out of its stomach, where it cannot brush away its adversary. Then the water-beetles are quite ready to eat each other should no better prey offer itself, while they ordinarily feed upon water-boatmen, water-gnats, and various larvæ. Yet, with all this mutual destruction, the creatures are not in the least afraid of each other, and a whirlwig larva will, for example, swim deliberately in front of a newt or a water-beetle, though its destruction is almost certain. I cannot but think that they do not look upon such a death as we do, and that the larger predacious creatures are to the smaller somewhat as disease and accidents are to ourselves—something which cannot be foreseen or avoided, and which has no terror until it actually comes to pass.

One more predacious insect, and we will conclude with two which are vegetarians, and which, though they find no food in their comrades of the pond, sometimes furnish it. I felt sure that in 'Mrs. Coates's Bath' the larva of at least one species of dragon-fly was likely to be found, and a part of a cast skin of a dragon-fly larva which I found in my net confirmed the theory. There were some rather large patches of duck-weed floating in the pond, which I thought were likely haunts for the creatures. Accordingly I put the net quietly under the duck-weed, drew it smartly with its edge against the floating plant, and at the very first dip secured three dragon-fly larvæ, all belonging to the genus Æshna, and being, indeed, the larvæ of *Æshna grandis,* one of our largest dragon-flies. I afterwards took more specimens, not only of that but of other species, including the Demoiselle, but the present creature is enough to act as a sample of the rest.

To my mind this is one of the most extraordinary beings that the world produces. There is no need of travelling to tropical countries for Nature's marvels. They are lavishly poured out at our feet, and we only have to recognise them. Its mode of progression is one that has lately been taken up as a new method of propelling steam-vessels, and its mode of seizing its prey displays a power of modification which very few structures attain. I took a number of these larvæ home, and watched their proceedings very attentively. They were well worth watching.

In the first place we will see how the creature propels itself. The body of the larva is long and tapering, rather larger in the middle, and ending in five horny spikes, which can be made to diverge from each other or can be pressed closely together, when they look like a single point. At the junction of these spikes is a circular aperture, large enough to receive an ordinary pin, and this aperture leads to a hollow space within the body. In this hollow are the gills, and respiration is carried on by means of the water which is drawn in and ejected through the orifice at the root of the horny spikes. When the creature is at rest the water is drawn in and out very quietly, and producing a gentle current that extends for several inches behind the larva. But when it desires to propel itself quickly the larva expels the water violently, and so, on the principle of the 'direct-action' machine, drives itself forcibly in the opposite direction. Thus the progress of the dragon-fly larva is necessarily a series of jerks, as some appreciable space of time is required in which the hollow can be filled with water. The nautilus, the common cuttle-fish and their kin, propel themselves in the same manner, which is exactly identical in principle with the flight of the rocket, and, in the creature called the Flying Squid, produces much the same effect.

If the larva be placed in a shallow and flat vessel, in which some very fine dark sand has been scattered, the whole process is rendered plainly visible. When the larva remains quietly in one place, the sand is gradually washed away in a direct line with the insect, leaving a track about a quarter of an inch wide, and some three inches in length. This track is very clear and well-defined near the insect, but becomes vague and broad in proportion to the distance from the larva. Now, if the larva be touched, a very different appearance is shown. The larva darts suddenly through the water, and, instead of the simple narrow track, a broad fan-shaped track is left, the water having been expelled with such force as to drive away the sand on both sides.

Its mode of eating is as strange as its progression. The lower lip, instead of being, as it mostly is, a mere

appendage to the mouth, is developed into a powerful
instrument of apprehension. It is greatly elongated,
being fully one-fourth as long as the entire insect. It
increases gradually in width from its junction with the
head to the end, which is armed with two short but
sharp jaws, curved and toothed in their interior edges.
It is furnished with two hinges, one at the junction of
the ' mask,' as it is called, and the other about half of
its length, so that it can lie flat against the breast, the
hinge descending as far as the base of the first pair of
legs, and the jaws lying exactly over the lower jaws of
the mouth. It is called the mask because its broad
end lies over the mouth and face of the insect so as to
conceal them. When the larva sees some creature
which it wishes to eat, it propels itself quietly beneath
its unsuspecting prey, turns over on its back, and, with
a sharp darting movement, seizes the unfortunate insect,
and holds it against the true jaws, by which it is soon
devoured.

The voracity of this larva is extraordinary, and it
seems capable of continually eating. As for my own
specimens, they were so voracious that at last I took
them out of the aquarium and put them into a vessel
of their own, supplying them with flies and other
insects. I found that, although they would eat blue-
bottles in lack of other food, they never seemed to like
them, although they would readily eat as many house-
flies as could be supplied to them. One day, thinking
that the formidable larva of the water-beetle was quite

able to hold its own, I put one of them in the same vessel. Next morning it was gone, and nothing was left of it but the two sickle-shaped jaws, which were lying at the bottom of the vessel. At last they took to eating each other, and I have now but one survivor, which, as may be expected, is a very large and fine specimen. It changed to the pupal state while in my possession, but is just as ravenous as it was when a larva, and as it will be when it becomes a dragon-fly.

Within this little pond are many species of caddis and several of May-flies—at least, of these insects in their preparatory condition. A really good collection of caddis-tubes can be procured from this spot, and I was rather surprised to find in it the curved and conical tubes of the Sericostoma, which are made of sand and tiny fragments of stone. May-fly larvæ also I found in tolerable plenty, and obtained them by the simple process of breaking up the mud of the bank, and catching them as they issued from their dwelling tunnels. These burrows are made in the soft muddy bank, and are shaped like the letter ⊃ laid horizontally, so that the inhabitants can pass in at one entrance and out at the other. As for larvæ of gnats and other flies, they simply swarm, and are present in such numbers that to give even a cursory description would take ten times the space that can be spared. The aquatic crustacea are in great numbers, and within the compass of that tiny pond may be procured enough specimens to give a laborious naturalist work for a year or two. Leeches, too, abound, and I was specially pleased to find several specimens of the Planaria, that curious flat-bodied annelid which is worthy of much examination.

In this and the preceding paper I have endeavoured to show the wonderful amount of interest which lies hidden in every object around us. Those who take up any branch of natural history pass straightway into a new world, and the more thoroughly do they enter into it the less do they complain of the narrowness of their field. I have intentionally taken two very narrow fields, namely, the living beings that are found 'Under the Bark,' and the creatures that live beneath the waters of a tiny pond measuring only three yards by four. And, so far from exhausting either the bark or the pond, I have given but the slightest and most sketchy account of both, choosing a few of the most conspicuous objects as examples of the rest, and leaving undescribed and even unmentioned hundreds of others every whit as interesting, but for which our limited space is insufficient.

MRS. COATES'S BATH.

A SUMMER WALK THROUGH AN ENGLISH LANE.

THERE are myriad spots in fair England most dear to the lover of nature, each having its peculiar attraction to the spirit of the spectator, and gladdening the soul of the poet or the artist with beauty as tender or majestic as can be found in most parts of this globe. But, of all beloved haunts, commend me to that which can be furnished by no other country on earth,—the real, dear, genuine, old-fashioned English Lane, with its banks of flowers, its little rippling streamlets, its shady hedgerows; its feathered trees, with their gnarled roots thrusting themselves out of the bank in strange knotty contortions, and occasionally making their appearance in the centre of the footpath, as if for the express purpose of flinging the heedless passenger on his nose; its charming freedom from any kind of regularity, its pleasant hum of busy insect wings, and its cheerful twitter of little birds. The woodbine flings its graceful masses of twining foliage and fragrant flowers over the hedgerows, and the odorous white blossoms of the wild clematis add their bright petals to vivify the scene. In some parts of the country this plant is called the traveller's joy, because it is supposed only to grow on the grounds of an honest man, and to wither straightway if he should fall into evil courses. Travellers, therefore, who come upon this flower may rejoice in their security, and place reliance upon the owner of the soil they tread.

Not in every part of England will you find the true unsophisticated lane—but there is no other country where you will find even its semblance. Some years since, a well-known American authoress paid her first visit to England, and was greatly charmed by the elucidation of a mystery which had long puzzled her while reading descriptions of English country life. Not until she had with her own eyes seen a genuine country lane could she understand how children could push themselves through the hedge after flowers, and so tumble into the ditch. Our painters have long discovered the value of lane scenery, and our truest poets have not been behindhand in painting with glowing words these uniquely lovely scenes of their native land.

At this time of the year, the exquisitely delicate tintings of the early leaves have passed away, and given place to a dark luxuriance of foliage, sobered here and there by the dried stalks of last year's vegetation, which underlie the light summer verdure, and are wonderfully effective in toning down the dappled greenage of the living leaves. To all who are capable of appreciating the many beauties of unrestrained nature the English lane is very dear; but to the field naturalist it derives an additional charm from the varied forms of life which swarm within its precincts. Every leaf is covered with a very world of minute beings; each bud and flower attracts thousands of happy and sportive existences within the sphere of its potent, though invisible perfume; and every plant is to creatures innumerable a cradle, a nursery, a banquet, and a home. The air is filled with the merry buzzing of insect wings that glitter in the sunbeams; the water teems with strange and weird-like forms; and even the apparently dull earth below the feet contains within its bosom beings as wonderfully mysterious in their structure and functions, though seldom, to our eyes, so lovely as the inhabitants of air. While we walk slowly through our country lane, let us pay a little attention to a few of the living hosts that are sure to cross our path.

There goes a great humble-bee, blundering along the flower-clad bank, with its steady, continuous drone, occasionally broken by a sharp, congratulatory buzz, as it alights on some untouched flower, and proceeds to rifle it of its sweet treasures. That is a maternal bee, hard at work as usual, gathering stores for her home, but taking very good care to give no intimation respecting her address.

The wiles of these insects are really astonishing. To find a humble-bee's nest is a common event enough; but to track the insect to her home is no such easy matter. She soon finds out that she is being watched, and tries to mislead her pursuer by artifices that would do credit to the cunningest fox that ever baffled a pack of hounds. She first tries to elude observation altogether, flies sharply to a little distance, settles on a plant, drops to the ground through the leaves, and either endeavours to lie hidden until the enemy has left the spot, or to crawl quietly away under the shelter of the foliage. It needs a practised eye to find the crafty insect as she crouches to the ground; and the best way is to rustle the herbage with a stick, and frighten her out of her hiding-place.

Off she goes in a great fume, humming and buzzing like a dozen bees, but never in the direction of her nest. Follow her up, and, finding that she cannot escape, she will change her tactics. She then tries to delude her pursuer into the notion that her nest is close at hand, and exhibits a vast amount of spurious anxiety about some little hole in the ground, about which she makes a great turmoil—crawling in, backing out, fluttering all round it, and making as great a fuss as if all her parental affections and household cares were centred in that little empty hollow.

Then, perhaps, she will pretend that she has not yet made her nest, and traverses the bank backward and forward as if she were seeking for a suitable locality, peering into every little crevice, scratching out a little soil here and there, and sometimes sitting quietly down for some moments as if quite fatigued. Turn your back for a minute, and Madam Drumbledore has vanished from the scene—slipped off quietly to her home in her own roundabout fashion.

Perhaps at another part of the lane, but certainly not within some distance from the spot where she was seen, the nest may be found, a mere insignificant hole in the bank, guarded in all probability by the roots of the neighbouring trees or bushes. Originally it was the home of a country mouse, deserted by the excavator, and squatted upon by the humble-bee. The inhabitants may be seen passing in and out at rather long intervals; and if the ear be applied to the aperture, a subdued kind of humming and buzzing is heard in the interior. There is no danger in this process, perilous though it may sound, for the big, heavy Drumbledore is among bees what the Newfoundland is among dogs, and seldom makes use of the formidable weapon with which Providence has armed her for 'defence, not defiance.' Many nests have I watched, and many have I opened, and never yet was stung by the humble-bee for my intrusion.

Dismissing therefore the fear of stings—for even if irritated, a humble-bee is so slow of wing that it cannot make the tiger-like charge of the wasp or the hornet, and can easily be captured or avoided—get out the long and strong bladed knife (which every observer ought to have in his pocket, together with string, a well-stocked pincushion, a supply of boxes, and a bottle half full of proof spirits of wine), and lay open the interior economy of the nest.

The spot in which the combs, if they can be so called, are placed, is always enlarged into a rudely globular apartment, in which are found a number of egg-shaped cells—not waxen and brittle like those of the hive bee, but brown in colour, and tough, soft, and of a leathery consistence. Neither are they arranged in a regular series, like the cells of the honey bee, wasp, or hornet, but are jumbled together without any apparent order, compacted into masses, and adhering to each other with tolerable firmness. Some of them contain honey of the sweetest and most fragrant character. Reader, beware that honey, or prepare for a headache and a giddiness for the next six or seven hours. Why the honey should have this effect, or whether it acts in the same manner upon all persons, I cannot say. I know, however, that in my own case, and in that of many others who have also had practical experience of this wild honey, the results have been almost identical. The remaining cells contain young humble-bees in every stage of their existence.

Interesting though the subject may be, I cannot within this limited space pursue it much further, although I should greatly like to say something of the economy of the sylvan home, and the wondrously modified structure of its inmates as they pass through their several phases of existence. Let me, however, very earnestly commend the humble-bee as an admirable subject for those who desire to study this portion of natural history for themselves. The creatures are of

large size, easily obtained; and in a single nest
examples may be found of the various states of this
bee, from the little white grub to the perfect insect of
either sex.

One curious story must yet be told of this subterranean home. Within the nest there are sometimes
found a few white grubs, clearly not those of the
humble-bee, as they are larger, straighter, and have a
row of spikes set around the larger end. If you manage
to remove the nest and put it into a box, so as to keep
its inmates prisoners, the mystery will be solved, in
time, by the appearance of some flies exceedingly resembling the humble-bee, but belonging to a different
order of insects—having only two wings instead of
four. This is one of the beautiful hovering flies, scientifically termed a volucella, the young of which finds
its food within the nests of these bees. The humble-
bees are quite aware of the injury to their community
which results from the intrusion of the volucella, and
are extremely vigilant in their watch to prevent its
intrusion. But the intruder is so like the insect into
whose house it hopes to make its way, that the two
can hardly be discerned from each other at a little distance; and so the volucella contrives to take advantage
of an unguarded moment, slips by the sentries, and
deposits her eggs. Having once succeeded in performing this feat, she cares no longer for her own safety,
but walks boldly out of the nest as if she had a perfect

right of passage. I have sometimes taken four or five of these grubs out of a single nest.

A near relation of this dipterous Paul Pry may often be found about the blackberries while they are in blossom. It is remarkable for the curious fact that the basal half of its abdomen is so transparent as to permit the colour of the leaves or petals to be seen through it. One of these flies, now before me, is so extremely transparent that when I place it on the paper on which I am writing the ink-marks can be seen through its substance, though not so clearly as to be readable, owing probably to the convexity of the abdomen. There are several British flies whose bodies are only semi-opaque, but there is none that can compare with the present example in the almost crystalline pellucidity of its structure. In consequence of this peculiarity it is called *Volucella translucens.*

A few paragraphs are now due to those much-dreaded insects, the wasp and the hornet, both of which may be found within the compass of our English Lane.

There are several kinds of British wasps, all very much alike in general appearance, but recognizable to the entomological eye by sundry slight, but legible marks. Some of these insects suspend their nests from trees, but the commonest species follow the example of the humble-bee, and choose a subterranean abode. Suppose now that we lay siege to a wasp's nest, as we have lately done to that of the humble-bee. 'Ware stings here, for there is no creature more irritable than your wasp, and it is by no means safe to go within hailing distance of a large nest. Even the exterior of their habitation presents a very different aspect to that of the humble-bee.

It is a busy scene. Around the entrance are crowding hundreds of yellow and black striped armed warriors, like the Pontifical guard on a small scale—some leaving the nest, and others hovering around for a few moments before entering, as if to inquire if all is well. You need not listen at the door of the establishment, for the humming buzz is quite audible, and the waspish temper is proverbial. I saw one nest whose inhabitants used to worry passengers to a great degree, and even attacked horses, and stung one poor animal so severely that it died from the effects of its many wounds. Sometimes a poor field-mouse, overtaken by a storm, runs into the apparently empty hole for shelter, but soon comes running out again, so covered with wasps that it looks like a yellow ball as it rolls down the bank beset with its angry foes. One of my friends, who saw a mouse thus assailed, calculated that from twenty to thirty wasps were at one time on the unfortunate mouse.

The nest of the wasp ought to be very carefully removed, so that its structure may be studied. If the nests of the hive-bee, the humble-bee, and the wasp be compared, they will be found to be made after three different fashions.

The combs of the hive-bee, are, as is well known, made of wax, secreted in certain little pockets situate in the abdomen. The edges of the cells are strengthened

with a kind of cement obtained from various trees, and
their shape is that of hexagonal or six-sided tubes, set
closely against each other, and practically carrying out
the interesting problem of giving the largest amount of
space with the smallest expenditure of material and
labour.   On the other hand, the cells of the humble-bee
are oval, and without any attempt at regular arrangement.   The walls of the cell are tough and leathery;
and, when subjected to the microscope, their structure is
resolvable into a number of regular silken fibres, crossing each other in a kind of meshless network, and
agglutinated together by some other substance.   But
the cell of the wasp is of a very different character from
both, and is composed of different snbstances.

The wasp makes his nest of veritable paper—not
quite so white or so fine as that employed in the printing of this book, but paper nevertheless. and made of
vegetable fibre, torn to shreds, pulped in water, and then
spread into sheets and dried.   Any one may see the
insect hard at work at its natural paper-mill.   Go to any
old post or decaying tree, and there may be seen the
wasps in full energy employed most zealously upon their
work.   Look at them closely—for they will allow themselves to be watched while thus occupied—and you will
soon see the process in its earlier stages.   With its
strong jaws the wasp bites away fibre after fibre of the
decaying wood, and continues to select a sufficient number to make up into a little bundle.   It is very
fastidious about the quality of the fibres, and rejects

almost as many as it retains. When it has obtained a sufficient load, it begins to champ and gnaw the fibres very diligently, moistening them at the same time with a drop of fluid, and being evidently absorbed in its work. Off it flies to its nest: but as we cannot see it there, we must take up a bit of wasp comb and discover how it builds up the cells.

On examination, we find that the walls of each cell are composed of this woody pulp, laid in regular strata, which are easily perceptible by the aid of a pocket magnifier. The walls are very flimsy, and cannot hold liquid; but as the English wasps make no honey, and store no food, this is of no consequence. The combs are arranged in regular layers, one above the other, each layer having all the open ends of the cells downward, and the closed ends forming a floor on which the insects can walk while traversing the space between the combs for the purpose of feeding the young grubs. Each layer is supported by a number of little pillars, about a third of an inch in length, made of the same *papier maché* substance as the cells, very much more solid and compact; and here and there a pillar is made very thick where the comb requires to be strengthened. On examination most of the cells will be found to be inhabited by white grubs, in every stage of growth. Many of the cells will be covered with white silken convex roofs, through which the black eyes of the future wasps often appear. The cells are not quite parallel with each other, but radiate slightly from the centre of each comb towards the circumference. The whole nest, containing some six or seven layers of comb, is enveloped in a kind of outer case, composed of the same paper-like substance as the cells, but of a much coarser consistence, and laid on in large loose flakes, in which the semicircular sweep of the wasp's head leaves its marks.

I have seen a very curious little wasp's nest taken from the neighbourhood of Balaclava during the Crimean war. All the stray wood was picked up and used for fuel, so that the wasps were deprived of their ordinary material. They soon, however, found a simple substitute, and made their nests of the blue and white cartridge-paper that is strewn in such quantities on a battle-field.

The wasp, although it makes no honey, is very fond of eating it, and is always allured towards any sweet substances with the same instinctive force which attracts the school-boy to the toffee-shop, or the infant to the sugar-basin. Ripe fruits are a great banquet to this marauder, who prefers the peaches, plums, and apricots to any other diet, and always chooses the juiciest and best flavoured upon the trees. But it is carnivorous also, and is a sad enemy to flies, to whom it is as deadly a foe as a winged spider would be. But here a poetical justice often overtakes the spoiler, for the hornet, shaped like himself, but just as much bigger, stronger, and fiercer as a tiger excels a leopard in these qualities, is particularly fond of wasps, and may be seen prowling about their haunts, sweeping upon them with a rush like that of the falcon, and carrying them off with ease, despite their wings and stings. It is rather curious that the hornet will not eat the head or the abdomen of the wasp, but shears them off with its strong jaws, bites off the wings and legs, and then crunches up the remainder just as we eat a radish. Sometimes the hornet flies to the branch of a neighbouring tree with the poor wasp in his jaws, and there slinging himself by one foot, he employs the remaining limbs in holding and arranging his victim to his satisfaction.

Here comes whirring along, rich in flashing green and glittering wings, the great dragon-fly, acknowledged tyrant of the air. He is arrowy-swift of wing, and there are few insects that can escape him. He cares little for birds, the general enemies of the insect tribe, for even the swift or the swallow cannot catch him, unless they come upon him unawares. See how he darts here and there, sometimes backing, by suddenly reversing the action of his wings with a sound like the ruffling of a small silken flag, and ever and anon pouncing upon some unfortunate insect as it flies along. Not even the broad-winged butterfly, with its erratic flight, can escape this dragon insect, although it gives him a hard chase. I have seen the poor butterfly dodge about like a startled snipe, or a coursed hare, in hope of escaping its terrible enemy; but all in vain. After two or three turns the dragon-fly succeeded in closing with its prey, and bore it unresisting through the air. As he flew along, wing after wing of the butterfly dropped from his mouth, and fluttered slowly through the air. For the dragon-fly was hungry, and was making the most of his time by devouring one victim while looking out for another.

If you want to observe him nearer, you can do so easily enough. Wait until he flies in your direction, and meet him with a firm sweep of the net. Down with the net on the ground, and seize the fierce creature, for he is biting his way through the gauze at a wonderful rate.

Take him out by his wings, and don't be afraid; for he cannot sting, though he is popularly called 'horse-stinger' by the rustics. Turn him on his back, and see how quickly and deeply he breathes, and how wonderfully the body is formed to permit of respiration. Pray do not think he is frightened. Not in the least; and we will prove it. Under the influence of terror, no creature will eat. But just take that fly out of the spider's web, and hold it to our dragon's mouth! See! he crunches it up in a moment; his mouth opens four ways at once; two pair of jaws and one pair of horny lips close on the fly and he is gone, with a snap and a bite, like a mutton chop down a Newfoundland dog's throat. Then you may give him the spider, and he will eat that too. He is very fond of spiders; and from certain observations, not yet published, I have a notion that spiders are almost necessary to these creatures, under certain circumstances. Try him with a beetle. Down it goes, but not so rapidly, the hard wing-cases having to be removed—which is done very adroitly and without pause. Will he eat a wasp? Certainly he will; and though I never tried him with a hornet, they being unchancy insects to hold while one hand is otherwise engaged, I have little doubt but that the hornet would soon disappear into the same receptacle with the other insects. You may go on catching insects for him as long as you like, and he will go on eating them, having no apparent limit to capacity. I once gave a dragon-fly thirty-seven large flies and four long-legged spiders, and only ceased because I was tired of catching before he was tired of eating.

Having admired him sufficiently, let him go. Off he darts to a branch, sits down for a moment, shakes his wings, as if to assure himself that they are fit for service, and then flashes off, as cruel and as voracious as if he had been fasting for a week. He does not spend all his time after this fashion, though he was always a predacious creature. His first few years were passed in the water, where he lurked under the banks, and chased the aquatic insects as fiercely as, when he got his wings, he pursued the inhabitants of air. In fact, the only change in him is that he was first a crocodile and then a dragon.

What beautiful butterflies, too, flit through our lane, varying with the time of day and the season of the year. There is the magnificent peacock butterfly, with its glorious 'eyes' upon the wings, like the spots on a peacock's train; the atalanta, or scarlet admiral, with its black wings edged with azure and crossed with a broad scarlet band; and possibly even the white admiral, the elegant camilla, may come flying along with that easy sweep of wing and that exceeding grace of movement which have earned for this insect the name of her who could skim over the waves without sinking, or over the ears of corn without bending their heads. Let us catch the peacock, just to look at the under-surface of the eyes. What a singular change. Instead of the varied colours which bedeck the upper surface, the whole of the under wing is deep brownish black, mottled and streaked by darker hues. Why is this? For a very sufficient reason, *i.e.*, to prevent the brilliant insect from being betrayed by its bright plumage. If alarmed, it instantly flies to some dark object, such as a tree trunk, closes its wings over its back, so that merely the dark under-surface is visible, and looks just like a dead leaf, or a strip of loose bark. When it was a caterpillar this was a curious creature, actually living on the stinging-nettle, and being itself covered with an array of handspikes curious to behold. I have bred hundreds of them and other butterflies from their earliest stages, and always found it to be the surest way of obtaining perfect specimens. Only, they must be liberally supplied with fresh food, or when they emerge from the chrysalis they are small and stunted.

Euphrosyne shakes her dappled wings from yonder thistle-top, the light sparkling from her silver jewelled plumes as she gently waves them in the sunbeams; over the grass-tops flits the exquisite Azure Blue, looking like a little bit of sky, bedropped with stars, that has come down on earth to gladden our eyes with its delicate beauty; and the bright Copperwing glides before us in glowing refulgence, as if its wings were veritably made of burnished gold. Moths, too, are not wanting, for the mottled Currant Moth flutters in and out of the hedge, displaying the rich cream and chestnut of its plumage. The Burnet Moth comes uneasily along with errant flight, pausing now and then long enough to show its green velvet coat, faced and trimmed with scarlet; and the swift Humming-bird Moth, agile as its feathered prototype, darts through the branches, poises itself on whirring wings before a flower, plunges its long proboscis into the nectary, and, taking some sudden alarm, is off like a lightning-flash.

Then there are the common but very beautiful tortoiseshell butterflies; the Janira, with its rich mottled brown plumage, and a host of others. If I could only be allowed the whole of this book for a description of a single insect, I might hope to do partial justice to the subject; but, under the present circumstances, we can only take a casual glance at each creature. White butterflies, of course, are flitting about everywhere. These may be destroyed mercilessly, or rather, mercifully. For pretty and harmless as they look, and as they indeed are, they are the parents of those horrid black, yellow, and green caterpillars that devastate our cabbage-gardens, and injure the temper of a hurried cook.   I say mercifully killed—because the caterpillar *must* be killed before we can eat the plant; and it is surely more merciful to kill one butterfly than some sixty or seventy of its future offspring.

The handsome Scarlet Hopper comes skipping and jumping so actively about the leaves that to catch it is no easy matter.   This pretty creature, with its scarlet and black clothing, is a near relative of the insect that forms the ' cuckoo spit,' so destructive to our garden plants.

Let us go a little farther down the Lane, towards that patch of bare sandy ground, and find out something about those bright blue flies that are dashing about it so vigorously.   See how they alight on the tawny soil, and how fast they run over its surface !   Now we see that they are not flies at all, but beetles, albeit they take to flight as readily, and are as active on the wing, as the blue-bottle flies, which they so closely resemble while in the air.   Catch one of them in the net as it flies along, and examine it.   What a pleasant perfume issues from its body !   Surely it must have been feeding on roses and verbenas.   No, it is a totally carnivorous insect, and rapacious to boot, and the agreeable scent is part of the mystery of its nature.   There is another beetle, very much larger, being at least ten times its size, called the Musk Beetle, which possesses a powerful rosy perfume, and, curious enough, is coloured after the same beautiful fashion.   Our little lively friend is called the Tiger Beetle ; and well does it deserve the name, for a winged tiger would not be more destructive among beasts than is the tiger-beetle among insects. What enormous projecting eyes it has ('the better to see you with, my dear,') and what long and powerful fangs ('the better to eat you with'). How firmly it is clad in bright and shining mail. deep, steely blue below, and green bedropped with gold and crimson above. Just look at its wing-cases through the pocket-magnifier, and see what a wondrously magnificent creature it is. Solomon was not robed half so gloriously as the lilies, nor were the Nepaulese princes half so gorgeously be-gemmed as this little beetle. Take it home; put it under the inch-power of the microscope, concentrate the light upon it with the condenser, and then say whether the jewelled beauties of Aladdin's palace could compare with the dazzling radiance of our little tiger-beetle! Fancy a few square yards of golden network set closely with emeralds, sapphires, and rubies as large as hazel nuts, and with diamonds as big and of more fiery splendour than the Koh-i-noor; illuminate them with the electric light, and you will then have some idea of the raiment with which God clothes even the smallest of his creatures. None can have the least conception of the hidden magnificence of the every-day objects around them except those who have studied them with a true and observant eye, and a sympathising and loving heart; and none but these can form so exalted an idea of the glories of a future life, which the earthly eye of man cannot see, nor his heart even conceive.

Now let us turn a glance towards the little streamlet caused by the drainage of the neighbouring fields, which has been quietly wending through its rushy path by our sides. Look steadily at every part of it, and the water becomes as thickly peopled as the land and air. On every side abundant living creatures are seen passing through the waters, some slowly, others rapidly, while many ascend from the bed of the stream, come for a moment to the surface, and dive away out of sight. The Water Boatman is very fond of that pursuit. A queer-looking creature is he, as he lies on his back in the water, his body shaped just like the hull of a ship, and his two long legs extended like oars on each side, and used after the same fashion. Catch him in the net, and look at him nearer—only take care of fingers ; for although the boatman cannot bite, he has a strong and sharp proboscis, and if carelessly held will startle his captor by inflicting a rather painful wound. Under his hard shelly wing-cases he has a beautiful pair of large membranous wings ; and at nightfall he leaves the water and takes to the air, mostly on some matrimonial business. At dawn he returns to the water, letting himself drop, with closed wings, from a great height. Sometimes the poor boatman falls into a sad error, and, mistaking glass for water, drops upon a greenhouse or a skylight, and kills himself with the shock.

In the more rapid and clearer parts of the stream, the Fresh-Water Shrimp may be seen driving itself through the familiar element by a series of jerks; now letting itself be carried down the stream for a foot or two, now recovering its position with a few strokes, and now scuttling aside in a desperate fuss, and hiding under a stone.

Perhaps, if we are very lucky, we may find one of the Water Spiders, and its wonderful nest, made exactly on the same principle as the diving-bell. The creature makes a silken bag under water, attached to some plant to keep it in its place. She then comes to the surface, gathers up a bubble of air, dives with it into her nest, lets it loose there, and returns to the surface for another supply. Each successive bubble displaces an equal amount of water, and in a short time the strange little architect has got a submerged palace, in which she can live as safely as on land. But, just now, we shall have to use our eyes very carefully to see the spider in her house, for the present rage for aquatic and marine vivaria has set a price on the head of the water-spider, and the country is ransacked by speculators to such an extent that where fifty such spiders might have been found in a single stream scarcely one can now be discovered. It is a cruel thing to take the poor spider away from her natural home and put her in an aquarium. She always dies soon; for although she may spin her wondrous house she cannot find her proper food, and is sure, before long, to succumb to her altered fortunes.

If we poke away the mud at the side of the stream, we shall probably come on some of the curious larvæ or grubs of the dragon-fly, which has already been mentioned, and may observe the funny way in which he propels himself forward by squirting water backward, having within him a 'direct action' propeller. Then his 'mask' is worthy of an examination. See how cleverly it is jointed to fit over his face, and how the formidable jaws at its extremity lie close to the head, so as to excite no alarm. Then see him dart out this mask to its full extent, snap up a passing insect in his jaws, and carry it to his mouth as an elephant picks up an apple and puts it into its mouth. If you like, you may take him home and keep him in a vessel of water, only he wants so much food that he is almost more trouble than he is worth—unless you have some special reason for watching his habits. He will eat little fish largely; and if you stock your water-vessel with young fry, this voracious creature will soon finish them.

Here comes from the bed of the brooklet the acknowledged tyrant of the waters, the great Water Beetle. He is so big that no insect can overcome him —so securely mailed that no insect, except his own kind, can hurt him—so swift that no aquatic insect can escape him; and so voracious that no amount of food seems to satisfy him. Even the dragon-fly grub had better keep out of his way; for he would soon be treated with poetical justice, and suffer the same fate he had so often inflicted upon others. Catch him, and hold him safely—taking care of your hands, for he can cut clean through the top of your finger if he gets a fair bite at it—and see how wonderfully he is made. The two fore-feet are swollen into circular cushion-like pads, which, if examined under the microscope, are seen to be set close with suckers of various sizes. Then his spiracles, or breathing-holes, are arranged under his convex wing-covers, which fit so truly against each other that they enclose a large quantity of air, which the beetle respires while he is below the water. It is to obtain a fresh supply of air that he comes to the surface, cocks up his tail, and dives perpendicularly downwards.

It will be of no use to take him home, for he must be kept by himself, or he would kill all other inhabitants of the aquarium in a few days; and he is so voracious that he requires nearly as much looking after as a young nestling. He will not even endure the company of his own species; and if another water-beetle be introduced, they will fight to the death. Even if you give him a companion of the gentler sex, his character loses none of its fierceness; and the two not only begin matrimony with a little aversion, but end it with the same—the conqueror always killing and eating the vanquished. How often have I not warned recent possessors of aquaria to beware the water-beetle; and how often have I not witnessed their despair at the death of all their stock, and the escape of the murderer, who has just taken to his wings and flown out of the window! The grub of the water-beetle is also aquatic, and one of the most repulsive and diabolical-looking creatures in existence. It is a great, fat, brown grub, as long as your finger, with a round, cruel-looking head, and a pair of huge crooked jaws, like two sickle-blades set on the head. Its voracity is wonderful. If you put one of these grubs in a water-tank, and then a piece of meat into the water, the grub seizes on it with its jaws, and keeps its hold for hours at a time. Should two or three of the water-beetle grubs be in the same vessel they will seize upon the meat; and if pushed with a stick will revolve like a wheel, all their heads being fastened to the meat, and all their tails radiating outwards.

Now let us walk on towards that shady pool through which our streamlet flows, and which has been deepened and embanked, and set round with trees, and guarded with stakes, to make it a fit drinking-place for the cattle. Nothing for a time is perceptible in the water, except those multitudinous little beings which are just large enough to be visible to the naked eye, and which are always playing through the water like motes in a sunbeam. Presently a small 'pop' is heard, a bubble is seen breaking, and just below the spot where the rippling circles arise, an indistinct gleam of orange and scarlet is seen through the disturbed water. Watch the spot carefully, and presently the same waving orange may be seen coming up from below, and assuming the form of a lizard-like reptile, some five inches long, with four legs, a well-developed tail, flattened at the sides to aid the creature in swimming—a beautiful crimson-stained and undulating crest, extending from the head along the back, and waving with every movement in the most elegant and graceful manner. This is the Triton, or common Water Newt, otherwise known by the name of the Eft, Effet, and Evat. He mostly lives in the water, and can exist for some minutes without needing to take breath. Every now and then, however, he must come to the surface to take a fresh supply of air, and in so doing he makes that odd little popping sound which we just now heard. He does not always wear that beautiful coat, for, like many of the birds, he only puts on his fine clothes during the matrimonial season, and for the rest of the year is clad as soberly as his mate.

Here he comes again; so slip the net under him quickly, and fish him out. Do not be afraid of him— he is one of the most harmless of beings, albeit he is popularly reported by rustics to spit fire, and to kill cows, and to bite pieces out of people's arms, and to sting like a viper, together with various other ill qualities; just as if he combined in his innocent five inches of dark tawny black and orange spotted belly all the demerits of the Dragon of Wantley, the rattle-snake, and the snapping turtle. Indeed, they could not display more fear of either member of that redoubtable trio than they exhibit when you pick up a newt and bring it towards them. As for yourself, your impunity will be set down, not to the harmlessness of the newt, but to some unholy compact with the powers of darkness. However, undeterred by such fears, let us examine the pretty creature more closely. Being out of the water, his beautiful crest is hardly visible, lying loosely along the back like the folded wings of a bat. See what lovely eyes the creature has, gleaming like fiery topaz, and unrivalled except by the eye of the toad. Put him down on the grass, and see how nimbly he runs to the water, and how he darts off with a powerful wriggle of his tail as soon as he finds himself safe. You will not catch him again very easily, for he has got a fright, and will take very good care of himself. If you look very carefully upon the slender leaves of the aquatic plant, you will probably find here and there a delicate, translucent, oval-shaped object, shorter, but thicker, than a grain of rice, but with the leaf curiously knotted over it. This is the egg of the newt, the tying of the leaf being performed by the forelegs—and a wonderful operation to see. There is much more interesting history respecting this pretty reptile, but other creatures are awaiting us, and we must pass onwards. I may just mention here that Mr. Knapp, in his 'Journal of a Naturalist,' remarks that he has seen the newts curiously encumbered with little bivalve shells on which they have trodden, and which have closed upon their unfortunate feet.

•　　　•　　　•　　　•　　　•　　　•

Evening is now drawing on; we retrace our steps through the Lane, and a new set of beings have come forth. The busy hum of insects has ceased, and the air is almost silent, except for certain odd squeaking noises that reach our ears. They cannot be made by a mouse, for they are too shrill; and besides, they come from above, and not from the earth. True, I can exactly imitate the sound by rubbing two keys together, but as keys don't go flying through the air, it will be safer to attribute the sound to the Bats, which are just coming abroad to begin their evening's work of gnat-hunting. Here they come! One, two, three—ten, twenty of them, fluttering with erratic but rapid wing like black butterflies against the darkening sky, and filling the air with their tiny shrieks, that pierce the ear like audible needles. You may catch a bat easily enough by sweeping the net sharply across its course as it flies down the Lane: only be chary of handling it, for its coat is always full of parasites which are by no means pleasant to look at. Put it on the ground, and see how beautifully its long membranous ears are formed, and into what graceful curves they are thrown, as the creature shuffles over the ground with that ludicrously awkward hobble which always reminds one of jumping in sacks. See, it spreads its wings and tries to fly, but only succeeds in tumbling on its nose. At last it scrambles to the top of the path, flings itself into the ditch—and so obtains sufficient impulse for its wings, and goes gladly off through the air.

Look under the hedge, and see how that leaf is moving along the ground, as if by magic. Stoop very gently, and you will see that a Worm has got hold of its point, and is dragging it towards his hole. I dare say that many who read this account will have seen in their gardens certain leaves sticking with their points in the ground, without knowing how they came in such a position. I did not know until last year, when I saw the whole process.

At dusk the worms begin to crawl out of their houses to hold friendly converse with their neighbours, and to survey the country. They never come entirely out of the hole, always leaving a joint or two within the aperture, by means of which they can retreat in a moment if alarmed. If you suddenly jerk a worm out of its hole it is quite at a loss; and even if you replace it by the former habitation it cannot find its old home, but is perforce obliged to make another. Watching these creatures is by no means an easy task, as they hate light, and seldom appear out of their holes except in the dusk, so that it is necessary to come quite close before they can be seen at all; and a lantern cannot be employed, as its glare would at once send them back into the darkness of their homes. The head of the worm is gently thrust into the air, the body follows, and then the creature begins to peer about in various directions, extending and contracting its body with great ease and rapidity. Presently it comes across a fallen leaf, pokes it about for a minute or two, seizes it by the point, and draws it to its home, always managing to hold it in such a way that when the leaf is dragged into the ground it is partially curled up. The worms

will take almost any leaf, though they evince strong
partiality when they have a choice.   Laurel, and other
evergreen leaves, they care little about, though, in de-
fault of others, they will use one of these.   But if a
lilac leaf be laid on the ground, the worm is sure to
find it out, will reject the laurel in favour of the lilac,
and draw the latter homewards.   The great favourite,
however, seems to be the primrose leaf, for which the
worm will desert any other plant.   It is curious to see
how long worms can make themselves when they want
to reach a leaf at a distance, and how thread-like they
then become.

In order to experiment upon them, I have laid
leaves with their stalks towards the hole, and always
found that in such cases the worm would feel its way
along the edge of the leaf, get hold of the point, twist
it round so swiftly that the eye can hardly follow the
movement, and then whisk it off homewards as if it
were moved by a spring.   No doubt, if we could discover some means of investigating them, we should find
the habits of the worms as interesting as those of the
insects.

Now the dew is collecting rapidly on the leaves,
and out come the Snail and Slug tribes from their
hiding-places.   Evening is the time for shell collectors,
as the lantern beams penetrate the dark recesses of
foliage, and bring out in bright relief the polished shells
as they move among the herbage.   Among the chief
favourites of the juvenile mind may be reckoned the

pretty Belted Snail, with its rich golden shell, banded
with deep mahogany or soft orange, as the case may be;
the Stone Snail, with its boldly-keeled shell seeming
as if it had been pinched along its length and then
squeezed flat; and the elegant little Clausilia, with its
long pointed shell looking as if made of flattened hairs.
Great black and grey slugs are now seen in every direction, descending the trees, gliding over the grass, and
showing themselves so plentifully that their hiding-
places during the day-time must be very cleverly
chosen to conceal such numbers of great fat creeping
things, which would be mercilessly extirpated if discovered.   Look now in the streamlet, and even by this
fading light other molluscs are seen inhabiting its
waters.   There is the Limnea, one of the commonest
aquatic shells, gliding about the water-plants, or floating down the stream on its back, having hollowed its
large foot so as to form it into a boat.   Sometimes a
whole fleet of these living vessels may be seen on their
voyage; and I have often known a continuous stream
of them of upwards of a hundred yards in length.
Throw a stone into the water, and down go all the boats
to the bottom.   There is another kind of limnea with
a wide trumpet-shaped mouth, nearly as common as the
former.   Towards the bed of the streamlet are seen
some flattened shells called by the name of Planorbis,
and looking something like the fossil ammonites with
which we are so familiar.

Out come the tribes of night-loving insects, heralded

by the swift moths as they shoot along the hedge-side, keeping low under cover of the shade, and flying so swiftly that they easily elude the stroke of the net unless held by a skilful hand. Strangest of all these moths is the Ghost Swift, appearing and vanishing as mysteriously as the most redoubtable ghost ever embalmed in tradition. One moment a white gleaming form is seen hovering in the air—another moment and it has gone. There it is again in precisely the same spot; so sweep the net with a smart side-stroke, and then examine the contents. See! there is a head of some flower, possibly a thistle, more probably one of the umbelliferous plants, and crouching in a corner of the net is our ghost-swift himself. I say himself, because the curious property of the ghost mentioned only belongs to the male insect. Look at the upper surface of the wings—they are shining white, while the under surface is golden brown. So, when the insect is hovering in the air, we only see the white wings shining through the duskiness, and when it settles on a flower or leaf, the brown under-surface is turned towards the spectator, and is quite invisible.

Can that be a snow-flake blown upon the hedge? Surely no snow can exist in this warm summer evening! Let us go closer, and examine it as it lies on the dark foliage. What a beautiful creature it is—a very flake of living snow—as white, as light, and more lovely. It is a most exquisite little moth, with its wings divided into five finger-like plumes, and covered with a deep fringe of the purest white. Even the body is clothed in snowy down ; and as the White Plume Moth flutters from one spot to another—for it never seems to take a long flight—it may easily be mistaken for a snow flake floating on the wind.

The great Hawk Moths now come dashing along, like the birds from whom they derive their name—and whom they resemble in no slight degree while on the wing—darting towards each flower and drinking its sweet contents by inserting their long trunks into its recesses, while they remain hovering in the air. You need not try to catch them, for the simplest plan to procure the most perfect specimens is by digging in autumn under any tree, wall, private hedge or paling, when you will find plenty of them in their chrysalis state ; and may procure the moths in absolute perfection by keeping them until the succeeding summer, when they will burst from their shells and come forth in their full beauty of unsullied plumage.

Now the Summer Chaffers and Dor Beetles come out of their retreats, bump against our face, or cling to our clothes, without seeming in the least discomposed by their sudden arrest. The former insect is a dreadfully destructive one, eating the grass-roots while it is a grub, and the trees when it is a perfect insect, and sometimes stripping tree after tree of its foliage. To the meditative saunterer at the evening hour it is an intolerable nuisance, irritating him with its buzzing hum even at his ear, sticking in his hair, dashing across his eyes, scratching his nose, and breaking up his train of thought in a lamentable manner. The Dor, with its beautiful purple-green body and helmeted head, is not nearly so tiresome—although it does occasionally thump us as it ' wheels its droning flight,' or startles us with its deep, heavy drone at our ears.

There is another beetle to whom we are more lenient. The Glowworm has just begun to light her blue-green lamp—a very Hero holding forth a torch to her Leander, who is flying above, and anxiously looking out for the welcome signal. See! down he comes with a wheel and furling of gauzy wings within their dark cases; and you may see Hero and Leander safely met. But how very odd! Hero is not one bit like Leander. She does not seem to belong even to the same class of beings, much less to the same species, as her lover. Leander is a long-bodied, wide-headed, brisk-looking beetle, with two ample pellucid wings enclosed in their protecting shields; whereas Hero is a flattish, slow, crawling, ordinary-looking, rather repulsive brown grub, with never a vestige of wing and nothing to recommend her to notice. Pick her up quietly, by gathering the leaf on which she sits; take her home; lay her on a bit of moist turf, and she will soon shine out for your gratification. But if you want her to be particularly resplendent, just pour a stream of oxygen gas through the vessel in which she is placed, and you will then see a blaze of natural illumination that can only be equalled by the many fire-flies of tropical regions.

I

Calm and quiet is the evening now, the sounds of labour are hushed and the bright songs of happy birds are stilled in sleep. But Nature has her vespers as well as her orisons ; the shrill cry of the bat and the deep humming of the circling beetle are psalms of praise as intelligible to sympathetic hearts as the sweet melody of feathered throats, or the pleasant sounds of busy insect wings. We who enjoy the blessed privilege of holding such sweet communion with Nature, and whose spirits are made capable of perceiving the Creator through the various forms in which He manifests His love, cannot do less than add our own heartfelt praises to those of all created things, with fervent thanks for the past, and loving trust for the future.

THE WOOD ANT.

WITHIN a reasonable walk of my house there is a small wood which affords opportunities of watching the habits of many creatures. It is planted upon more than one kind of soil, so that a plentiful supply of plants is found within its limits, and, as a necessary consequence, many insects live upon the plants; predacious insects come to eat their harmless kinsfolk, and birds come to eat both the cannibals and their victims.

This place is a favourite resort of the Wood Ants, which have built their fragile but enduring nests in many sheltered spots, and have driven their wonderful paths through almost every part of the wood. For some years I have passed many pleasant hours every summer among the trees, and found the day only too short for the many observations that came under my notice. The best way to take advantage of a wood is to set out with the intention of watching some particular creature, and to give up one's time exclusively to that single object; not failing, of course, to mark any point of interest that may present itself respecting other beings that may come within ken, and to jot it down in a note-book.

This insect, which may be known by its large size and reddish thorax, is one of the stingless ants, though it is quite as formidable an antagonist as those species which possess those sharp and envenomed darts. For though the wood-ant has no sting, it yet has a store of poison, and can use its venomous powers effectively, though in a more roundabout manner than is adopted by the sting-bearers.

It is a most fierce and determined creature, the sense of fear seeming to have been wholly omitted from its composition. It will attack anything and anybody without the least hesitation, and possesses all the courage without the cunning exhibited by the Lilliputians in their memorable attack on the Man Mountain. For a man is to the wood-ant not only a moving mountain, but a moving world ; and yet there is not a single ant that will not attack a man, if it fancies him to be in too close proximity to its residence.

Urged by some wonderful instinct, it makes at once for the nearest unprotected skin, bites fiercely with its sharp and calliper-shaped jaws, and simultaneously bind- ing its body so as to bring the tip of the abdomen to bear upon the wound, squirts a small drop of its poison into the cavity, producing for the time a sharp and painful smarting sensation. The pain, however, is very transient, although, at the moment it is inflicted, the pang is quite as severe as that inflicted by the sting of a wasp. Nor is this its only mode of attack. The wood- ant is able to eject this poisonous substance to some distance, and if a nest be broken open, and a bare hand placed within the aperture, it will be speedily covered with a thousand little dots of pungent fluid; and if the skin be very sensitive it will smart as though it had been plunged into a bunch of stinging nettles. The scent of this fluid is strongly acid, like highly concentrated vinegar, and even at a distance of a yard from the nest produces an unpleasant sensation in the throat and nostrils.

One of my friends, desirous of testing personally the peculiar scent, made a breach in the nest of the wood-ant, and put his face to the hole. Scarcely had he approached within three inches than he started back, vowing that the ants had stung him all over his chin, and could not for some time be convinced of his error.

This pungent liquid is acid in its nature, and when analysed is found to contain two kinds of acid, one peculiar to these insects and called 'Formic' acid, and the other the substance called 'Malic' acid, which gives to the juice of apples its peculiar flavour. Not only has it the scent of vinegar, but a very good substitute for that useful article is often made by steeping successive measures of the wood-ant in boiling water. The substance called chloroform owes its name to the similarity between its constituent elements and those of Formic acid. In chemical language, though not in chemical formula, Formic acid consists of two atoms of carbon, one atom of hydrogen, and two atoms of oxygen, while the composition of chloroform is two atoms of carbon, one atom of hydrogen, and three atoms of chlorine. It may be casually remarked that Formic acid can be produced by artificial means.

The nest of this insect is a wonderfully large structure, when the size of the tiny architects is taken into consideration, and the regularity with which their interior is parcelled out into chambers and galleries is not less surprising. It is made of little pieces of stick, dried leaves, broken stems of the dried fern, and always contains the berries of the mountain ash, if any tree of this kind should happen to be within a moderate distance. When I first observed the red berries amid the heap of leaves and sticks, I thought that they had fallen from a neighbouring tree and been accidentally blown upon the nest, but I have since found that every nest in the wood contains these berries, when a mountain ash was within forty or fifty yards. Their use I cannot imagine, as the ants do not carry them into the nest, but merely mix them with the dried substance of the exterior.

The materials of which the nest is composed are heaped quite loosely and apparently at random on each other. But if the nest be carefully examined, a certain order is to be detected, particularly in the entrances and galleries, which are made of long sticks rudely arranged across each other, so as to form a five-sided aperture. If a twig be brought to the nest, its destination is nearly sure to be at one of the many openings. Being desirous of ascertaining whether the ants would accept extraneous assistance, I broke off a little dried stick to the shape and size of those that were arranged about the aperture and laid it upon the others, so as to match them as nearly as possible. A posse of ants immediately came to look at the new addition, took hold of it with their jaws, and after making a trifling alteration, for form's sake, I suppose, lest I should be too conceited about my architectural skill, they allowed it to remain.

It is most interesting to watch the ants bringing materials for their home. If an ant finds a little piece of broken fern stem that is suitable for the outer wall, he picks it up by one end, holds it out straight before him as if he were smoking a very large cigar, and sets off briskly homewards. This mode of carrying his burden is evidently adopted for the convenience of steering it through the grass blades, fallen fern, and other impediments, which, trifling as they appear to human eyes, are by no means insignificant to the ants. I have even seen an ant carrying off a grub three times his own size, holding it in the same manner; the strength required for such a feat is truly enormous.

But when a heavier or a larger burden, such as a piece of stick, has to be transported, a different plan is adopted. Six or seven ants are detached for the work, and they set about it with a nicety of purpose that is really surprising. Grasping it with their jaws they gradually edge it onward in the right direction, one of their number always seeming to act as foreman, and taking hold of the end which seems to be the post of honour. As long as the ground is tolerably even the stick is dragged along without difficulty, and the foreman or 'ganger' cannot be distinguished from his fellows save by his position at the end of the stick. But when they get among broken ground, or if the stick should perchance fall into a crevice and carry its leaves with it, the ganger seldom touches the stick except to pull it into the proper direction, but runs ahead to reconnoitre the locality, then returns to the gang, and is all life and animation. I have seen the clever little creatures make a mistake, and get the stick into a labyrinth of broken ferns and twigs, through which they could by no means steer it, and then seen them carefully return by the same path until they were clear of the thicket, and choose another and a smoother road.

On one occasion I watched a gang of ants, six in number, that had jammed their burden so tightly under a fern stem that they could proceed no further. They immediately tried to extricate it, but were checked by an angular bend in the stick, which had hitched itself under the fern, and prevented it from being moved in either direction. Being curious to know how the ants would surmount the difficulty, and rather fancying that they would leave the stick and fetch another, I watched them for nearly two hours. They evidently had no intention of relinquishing their task; and after a vast amount of excitement, the ganger getting on the top of the stick and down again fifty times, they hauled the projecting extremity down by main force of numbers, dragged it from below the impediment, and, I suppose, got it safely home. The stick was a trifle more than two inches in length, about as thick as a stout crow-quill, and at one end had a knot and a sharp bend upwards. An idea of the strength exerted in the transportation of this burden may be formed by taking the comparative sizes of men and ants, magnifying the piece of stick into a tree trunk of corresponding dimensions, and setting six men to carry that trunk through a virgin forest, and over ravines and precipices, up mountains and down valleys, and lastly to the top of a building shaped something like the great pyramid, but much more lofty, the sides of which are formed of loose sticks and logs.

Nothing short of taking away the object of their labours seems to divert these industrious creatures from their work. I have laid large flies, little grubs, and other attractive articles of diet in their way, but they suffered them to remain unheeded, though, if unemployed on serious business, they would carry off such prey as soon as they saw it.

The wood ants seem to be acquainted with the leading principles of civilisation, their nest being the centre of a radiating system of roads, extending for a wonderful distance, and as permanent in their way as Watling Street, or any of the old Roman roads which now traverse our land. Mr. William Howitt tells me that he has watched one of these roads for more than twenty years, and found that on every fine day it was crowded with ants going off for plunder, or returning laden with spoils for the benefit of the community. Even on wet and cold days, when the ants, who are chilly beings, wisely stay at home, their roads are plainly perceptible, and are marked out by their freedom from bits of stick, leaves, etc., these having been removed by the insects as materials for their nest. It is always easy to find the nest by following up the road, and the right direction can be at once learned by following the course adopted by the laden insects.

The difference in the demeanour of those that are setting out in search of prey or materials and those that are returning home is most notable ; the former bustling along with a quick eager step, looking this way and that, running first to one side of the path and then to the other, interchanging rapid communications with their comrades, and altogether brisk and busy. But when they have succeeded in their object they march steadily homeward with a preoccupied demeanour, taking no notice of passing events, and being apparently absorbed in the one task of depositing their burden in its proper place.

The observer will do well while watching these insects not to sit or stand upon or very near one of their roads, for the ants have no idea of being pushed out of the old paths, and are summary and fierce in their revenge upon intruders.

As the ants pass and repass on their paths they hold

rapid communications with each other, mostly by means of their antennæ, which pat and stroke those of their gossip with surprising quickness, the whole transaction irresistibly reminding the observer of the Oriental method of conducting sales or barters by means of the hands. The antennæ, whose precise function is still rather obscure, are employed .not only for actual communication with other ants, but to ascertain whether a companion has passed over a certain spot. This peculiar instinct is mostly exercised among trees. The ant roads seem even to extend themselves to the summit of trees, being generally confined to one side of the trunk, and ramifying to the very tips of the leaves, as may be seen by means of a good field-glass. Ants may be seen passing and repassing upon the trees as briskly as upon the ground, and it is notable that when they get among the small branches an ant will not go where another has preceded it, making itself aware of the circumstance by the tapping of the antennæ upon the bark.

The object of this tree-haunting habit is twofold; firstly that the individual may obtain food for itself, and secondly that it may bring in subsistence for the community. Its own nourishment is chiefly obtained from the aphides which swarm on many trees, and which have the power of exuding a saccharine fluid from a pair of minute tubes near the extremity of the body. When the aphides are very plentiful the sweet juice falls on the leaves, and is popularly known as the

' honey-dew.' Both bees and ants are fond of honey-dew, which the former insect licks from the leaves with its brush-like tongue, the latter taking a more direct course, and lapping it as it exudes from the tubes. While on the leaves the ants are more than usually combative, and if the hand be placed near them, will tuck their tails under them, sit up like dogs begging, and flourish their antennæ in a manner which they doubtless think well adapted to frighten the disturbers of their peace. When, however, the angry insect finds that menace is ineffectual, and that it cannot alarm the foe, it settles the matter by dropping to the ground. If an ant-infested tree be suddenly struck with a stick, the ants tumble down in all directions, falling quite unconcernedly from a height of fourteen or fifteen feet, and rattling like hail upon the dried leaves at the foot of the tree. When they reach the ground they lie motionless for a moment, and then pick themselves up and run away as if nothing were the matter.

Though they instinctively spare the aphides (an instinct which every gardener cannot but wish to be suppressed), they are terrible foes to other insects, seizing them and dragging them into their nests most zealously. I once saw an unfortunate daddy long-legs (*Tipula*) caught in a gust of wind and blown upon a nest of the wood-ant. No sooner had the ill-fated insect touched the nest than it was surrounded by a host of ants, its legs seized by twenty pairs of jaws, its wings torn from their joints, and the still struggling body pushed and dragged along until it was finally pulled into the recesses of the nest. I have often tried the experiment of putting a large fly in their path, and always found their mode of procedure to be the same. They cluster round the fallen insect like flies round a lump of sugar, they seize upon its legs, they pull off its wings in a moment, and run away with the severed organs, four or five others following each fortunate captor, just like a brood of chickens after the one that has been lucky enough to pick up a piece of bread. They then attack the wingless body with ruthless violence, biting at it like a hungry cat at a slice of meat, or perhaps more like a herd of wolves at their prey. They soon deprive it of life, haul it to the nest, drag it up the side, and literally tumble it into one of the holes.

The interior of the wood-ant's nest, and the mode by which a view of it was obtained by the insertion of a sheet of plate-glass, are described in my 'Homes Without Hands.'

THE GREEN CRAB.

Of all the animated denizens of our sea-shores, there is perhaps none more generally familiar than the common green crab, or shore-crab as it is also popularly termed— the *Carcinus Mœnas* of naturalists. Whether at high or low water, at ebb or flow, hiding under overshadowing weeds or craftily sunk beneath the sand, this quaint, waddling, green-backed crustacean is to be found, equally active, and equally pugnacious. With the exception of children, who are always delighted with the odd manœuvres of the creature, people mostly look upon it with contempt, partly because it is too small to hurt them much, and partly because it is not particularly worth eating, having hardly anything inside its olive-green shell, and the little that there is not being well-flavoured. Yet beneath that unprepossessing exterior is concealed a vast fund of interest, and the visitor to the sea-side will find himself well repaid by watching the habits of our olive-coloured friend.

The best time and place for observing the green crab in the fulness of its energies is just before high tide. Just at the edge of the advancing waters, crabs rise out of the sand in all directions, like the warriors sprung from the dragon's teeth, and, as if to complete the analogy, each is supplied with defensive and offensive armour, and each is at mortal enmity with its companions.

As the waters roll towards the shore the crabs advance with the waves, ever hovering on the extreme verge, and hungrily watchful for their prey. The dashing waves tumble them over in a most unceremonious fashion, but without in the least disturbing their equanimity, and it is amusing to see how cleverly they guard themselves from being washed back into the sea by sticking their hooked legs into the sand, like animated grapnels.

Before watching the habits of the creature, just let us catch one, and examine the marvellous manner in which its form is adapted for the life which it leads.

The legs are so constructed that they permit their owner to move backwards, forwards, or sideways with equal ease, a capability which is of the greatest importance in procuring food, as well as in escaping from foes. The latter contingency is also beautifully provided for by the shape of the body, which is so formed as to enable the creature to burrow beneath the sand with singular rapidity, leaving scarcely a trace of its presence.

To watch the animal thus employed is an interesting sight. The crab half erects itself on its tail, and scoops up the sand with the edge, just as a child digs a hole with its wooden spade. If the sand is wet, three or four vigorous movements are sufficient to sink the crab below the surface, when the next wave washes a quantity of loose sand over the spot, and nearly obliterates the traces of the creature that is lurking below. A practised eye will, however, detect the concealed crab by means of the bubbles that issue from the sand in consequence of the air expressed from the system.

Here we may mention that the proper way to catch a crab without being bitten is to press the forefinger smartly on the middle of the back, and then to grasp the two side edges with the thumb and middle finger. The claws are thus forced to fold their joints, and their painful bite need not be feared.

Holding the crab in this manner, turn it over, and examine the wonderful manner in which the limbs are packed, and how admirably they accommodate themselves to the habits of the animal. The claws, when folded, exactly bring their extremities to the mouth, so that any food can be carried to the right place, and literally 'tucked in.' The mouth itself is an apparatus so complicated that it cannot be described further than as being a series of jaws and teeth, placed behind each other in regular succession, and opening like horizontal shears.

A creature that depends upon its own exertions to capture the active prey on which it feeds must necessarily be furnished with powerful eyes, which are capable of extending the faculty of vision over a very large field. These eyes are seen on the front margin of the crab, placed on footstalks, and having a peculiar nacreous lustre on their grey-brown surfaces. On examination

CRABS AT HOME.

with a good pocket lens, the eyes are seen to be compound, *i.e.* formed of a great number of facets, each possessing the power of vision, and all communicating with their common optic nerve. The delicate raised lines caused by the serried ranks of these compound eyes are the origin of the peculiar lustre just mentioned. It will be seen, too, that the visual portion of these organs passes partially round the footstalks, so that when the creature protrudes its eyes it can see objects on all sides with equal ease.

Now replace the crab in the water, and watch it as it exhibits the instinct which has been implanted in its being by its divine Creator.

Advancing with the flowing tide, and even remaining within a foot or two of the edge, the crab keeps its eager watch for food, and suffers few living things to pass without capturing them. The whole nature of the animal seems to be changed while it is seeking its prey. The timid, fearful demeanour which it assumes when taken at a disadvantage wholly vanishes, and the apparrently ungainly crab becomes full of life and spirit, active and fierce as the hungry leopard, and no les destructive among the smaller beings that frequent the same locality.

Now does it shew the ubiquitous advantages of its singular mode of progression. Let a tiny fish, a smaller crustacean or a soft mollusc, pass it within reasonable distance, and the crab darts at it with a tiger-like energy, and seldom fails to secure its prey. I have seen these crabs run after and catch the black flies that are so common upon the sand, and once saw a burrowing wasp (*Odynerus*) snapped up as it alighted on a bit of seaweed; and I have often seen bees thus caught as they were drinking the salt water. Everyone who has walked along a sandy shore at evening is familiar with the shrimp-like sand-hoppers or sand-skippers (*Talitrus*) that leap about with such untiring energy, and knows the difficulty of capturing one of these active creatures. Yet I have seen the green crabs give chase to the sand-hoppers, and pounce on them as cats on mice.

The method employed in their capture of all active animals is unique. As soon as the crab sees the intended prey it sits up for a moment, darts at the doomed being, and literally flings itself upon the victim, imprisoning it beneath the body and hemming it in by means of the legs, which make an impassable cage around it. One of the claws is then inserted under the body, and the prisoner picked daintily out as if by the thumb and finger. One claw then holds the prey, while the other pulls it to pieces and puts the morsels deftly into the mouth. The crab knows the value of time, and loses not a moment in disposing of its prey, tucking it into its voracious maw with amusing despatch, and looking out the while for a fresh victim. Once I saw a very large sand-hopper make its escape from its pursuer. It gained nothing, however, but a temporary release, for the crab instantly gave chase, secured, and ate it in a few moments.

Fierce and destructive as it may be, the green crab is itself a frequent victim to more powerful foes, and is often doomed, with poetical justice, to undergo the sufferings which it has inflicted upon other beings.[1] None are more terrible enemies than those of its own species, for the crab is an insatiate cannibal, devouring its own kindred without the slightest compunction. In all these cases, however, it is needful that the dimensions of slain and slayer should be very disproportionate, as the crab cares not to earn a meal through a fight.

I was lately witness to a very amusing episode, where a large and powerful crab caught sight of a tender little one, as it scuttled over the wet sand. Away started the giant in full chase, and away ran the pigmy, as if knowing that life and death hung on the issue of the race. In spite of the great disproportion in size, the superior activity of the smaller crab prevented its pursuer from gaining much ground, but at last its strength evidently began to fail, and I thought it must inevitably succumb to the terrible foe that pressed so fiercely on its footsteps. Suddenly, however, it darted under a stem of laminaria that was lying on the shore, gathered all its limbs under its shell, and there lay motionless. The pursuer was instantly baffled. It raised itself in the air, and surveyed the shore in all directions. Then it prowled about like a cat that has lost a mouse. It even

---

[1] Mr. Gosse mentions that he saw one crab, while eating another, seized by a larger crab and eaten in his turn. He did not seem to be sensible of the fact, but went on eating until he was entirely crushed.

then was cunning enough to turn over some bits of sea-weed that were lying on the shore, but never thought of searching under the thick stem of the laminaria. At last it gave up the pursuit, returned disconsolately to the sea, shovelled itself under the sand, and I saw it no more. Its intended victim then cautiously looked from its place of shelter, just protruded a claw, then a leg, then looked again, and at last came boldly forth and went off to catch something on its own account.

As a general rule, the larger the size of the crab the more bellicose is its disposition. The smaller specimens are usually discreet as well as valorous, and if surprised either run away as fast as they can, or burrow into the sand with all speed. But the great broad-shelled bully of the rocks has had his own way so long that his first impulse is always to show fight, and no sooner does he catch sight of foe than down goes his tail and up go his claws, and there he sits, defiantly ready for instant combat. It is as well to be cautious about handling such a champion, for he can strike with his claws as swiftly as a serpent darts its armed head, and should he miss his aim the clash of the bony weapons is distinctly audible.

Be it well understood that a bite from such a creature is no trifle, for the claws are enormously powerful, their tips are sharply toothed, and they hold like the jaws of a bull-dog.

Even this belligerent animal is ofttimes fain to retreat before a foe of greater powers, stronger weapons, and harder shell—namely, the edible crab, which figures on our tables, and is known among the seaside population as the punger. Fortunately, however, for our green friend, the punger mostly inhabits a different belt of water, being most commonly found among the rocks at low-water mark, whereas the green crab lives almost wholly above that elevation.

Many persons when walking along the shore have observed a curious series of little marks on the sand, set in rows of five or so in depth, and meandering in seemingly purposeless irregularities. At first the marks appear to be made without any order, but a little examination will show that the same group of marks is repeated at regular intervals. These are the foot-tracks of the green crab, and the distance between the parallel lines of marks denotes the size of the animal that made them.

Guided by these tracks, an experienced shore-haunter can often follow the crab to its place of concealment and bring it to light, whether it be buried in the sand, or lying under the shelter of pendent seaweeds. In attempting this feat, however, it is as well to be quite sure of the direction in which the crab has gone, so as not to be led away from, instead of towards, the hidden crustacean. This object can easily be attained by examining the shape of the marks, which are always larger at one end than the other, the larger extremities always pointing in the direction the crab has taken.

There is much more to be said of these creatures, but failing space will not permit of further description. Should, however, any reader of these lines suffer the annoyance of a wet day at the sea-side, he is hereby recommended to procure a waterproof garment, to betake himself to the shore as the tide is rising, and amuse himself by watching the crabs.

*   *   *   *

Since this article was printed, I have found that the green crab supplies food to rats. Along the cliffs of Margate great numbers of rats live, having taken up their abode in the crevices, which are above the reach of the waves. My attention was first called to them by seeing their footprints in the thin coating of sand which is to be found in such places, the sand having been blown there by the wind.

A friend and myself then set to work at these marks, and soon found that great numbers of rats inhabit the cliffs, and that they live almost entirely on the green crabs, whose broken shells lay plentifully in the crevices. After watching for some time without success, we at last managed to see the rats stealing out of their homes at dusk, and carrying off the crabs. It was very difficult to see the animals, for they are scarcely visible against the sand and seaweed, and are so wary that the least movement sends them scuttling off to their homes. Sometimes, when the crab happened to be a large one, the rat had no easy task, for the crab was quite as courageous and nearly as active as its foe, springing round so as to keep its front to the enemy, and guarding itself with its claws just as a boxer holds his arms when on guard.

I may mention that the popular idea of the green crab being unfit for food is entirely wrong. The flesh is nearly as good as that of the edible crab, but there is so little of it that the creature would command no sale in the market, and is therefore left undisturbed.

MEDUSA AND HER LOCKS.

ALONG the sandy shores at low water may be seen in the summer months numbers of round, flattish, gelatinous-looking bodies, scientifically called Medusæ, going popularly by the expressive though scarcely euphemious titles of slobs, slobbers, stingers, and stangers, and called jelly-fishes by the inland public, though the creatures are not fishes at all, and have no jelly in their composition.

As these medusæ lie on the beach they present anything but agreeable spectacles to the casual observer, and, as a general fact, rather excite disgust than admiration; and it is not until they are swimming in the free enjoyment of liberty that they are viewed with any degree of complacency by an unpractised eye. Yet even in their present helpless and apparently lifeless condition, sunken partially in the sand, and without a movement to show that animation still holds its place in the tissues, there is something worthy of observation, and is by no means devoid of interest.

In the first place be it noted that all the medusæ lie in their normal attitudes; and in spite of their apparently helpless nature, which causes them to be carried about almost at random by the waves or currents, they in so far bid defiance to the powers of the sea that they are not tossed about in all sorts of positions, as is usually the case with creatures that are thrown upon the beach, but die, like Cæsar, decently, with their mantles wrapped round them.

Looking closer at the medusa, the observer will find that the substance is by no means homogeneous, but that it is traversed by numerous veinings, something like the nervures of a leaf. These marks indicate the almost inconceivably delicate tissues of which the real animated portion of the creature is composed, and which form a network of cells that enclose a vast pro-portionate amount of sea-water. If, for example, a medusa weighing some three or four pounds be laid in the sun, the whole animal seems to evaporate, leaving in its place nothing but a little gathering of dry fibres, which hardly weigh as many grains as the original mass weighed pounds. The enclosed water has been examined by competent analysts, and has been found to differ in no perceptible degree from the water of the sea whence the animal was taken.

Though the cells appear at first sight to be disposed almost at random, a closer investigation will show that a regular arrangement prevails among them, and that they can all be referred to a legitimate organisation. So invariably is this the case that the shape and order of these cells afford valuable characteristics in the classification of these strange beings.

Just below the upper and convex surface may be seen four elliptical marks, arranged so as to form a kind of Maltese cross, and differently coloured in the various specimens, carmine, pink, or white. These show the attachments of the curious organisation by which food is taken into the system, and may be better examined by taking up the creature and looking at its under surface.

Now take one of the medusæ, choosing a specimen that lies near low-water mark, and place it in a tolerably large rock pool where the water is clear, and where it can be watched for some time without the interruption of the advancing tide.

The apparently inanimate mass straightway becomes instinct with life, its disc contracts in places, and successive undulations roll round its margin like the wind waves on a cornfield. By degrees the movements become more and more rhythmical, the creature begins to pulsate throughout its whole substance, and before very long it rights itself like a submerged life-boat, and passes slowly and gracefully through the water, throwing off a thousand iridescent tints from its surface, and trailing after it the appendages which form the Maltese cross above-mentioned, together with a vast array of delicate fibres that take their origin from the edge of the disc, or umbrella, as that wonderful organ is popularly called.

Words cannot express the exceeding beauty and grace of the medusa, as it slowly pulsates its way through the water, rotating, revolving, rising and sinking with slow and easy undulations, and its surface radiant with rich and changeful hues, like fragments of submarine rainbows. It is often possible, when the water is particularly clear, to stand at the extremity of a pier or jetty and watch the medusæ as they float past in long processions, carried along by the prevailing currents, but withal maintaining their position by the exertion of their will.

The reader is doubtlessly aware that the title of Medusa is given to these creatures on account of the trailing fibres that surround the disc, just as the snaky locks of the mythological heroine surrounded her dreadful visage. Many species deserve the name by reason of the exceeding venom of their tresses, which are every whit as terrible to a human being as if they were the veritable vipers of the ancient allegory.

Fortunately for ourselves, the generality of those medusæ which visit our shores are almost, if not wholly, harmless; but there are some species which are to be avoided as carefully as if each animal were a mass of angry wasps, and cannot safely be approached within a considerable distance. The most common of these venomous beings is the stinger or stanger, and it is to put sea-bathers on their guard that this article is written, with a sincere hope that none of its readers may meet with the ill fate of its author.

If the bather or shore wanderer should happen to see, either tossing upon the waves or thrown upon the beach, a loose, roundish mass of tawny membranes and fibres, something like a very large handful of lion's mane and silver paper, let him beware of the object, and, sacrificing curiosity to discretion, give it as wide a berth as possible. For this is the fearful stinger, scientifically called *Cyanea capillata*, the most plentiful and the most redoubtable of our venomous medusæ.

My first introduction to this creature was a very disastrous one, though I could but reflect afterwards that it might have been even more so. It took place as follows.

One morning towards the end of July, while swimming off the Margate coast, I saw at a distance something that looked like a patch of sand, occasionally visible, and occasionally covered, as it were, by the waves, which were then running high in consequence of a lengthened gale which had not long gone down. Knowing the coast pretty well, and thinking that no sand ought to be in such a locality, I swam towards the strange object, and had got within some eight or ten yards of it before finding that it was composed of animal substance. I naturally thought that it must be the refuse of some animal that had been thrown overboard, and swam away from it, not being anxious to come in contact with so unpleasant a substance.

While still approaching it I had noticed a slight tingling in the toes of the left foot, but as I invariably suffer from cramp in those regions while swimming, I took the 'pins-and-needles' sensation for a symptom of the accustomed cramp, and thought nothing of it. As I swam on, however, the tingling extended further and further, and began to feel very much like the sting of a nettle. Suddenly the truth flashed across me, and I made for shore as fast as I could.

On turning round for that purpose, I raised my right arm out of the water, and found that dozens of slender and transparent threads were hanging from it, and evidently still attached to the medusa, now some forty or fifty feet away. The filaments were slight and delicate as those of a spider's web, but there the similitude ceased, for each was armed with a myriad poisoned darts that worked their way into the tissues, and affected the nervous system like the stings of wasps.

Before I reached shore the pain had become fearfully severe, and on quitting the cool waves it was absolute torture. Wherever one of the multitudinous threads had come in contact with the skin there appeared a light scarlet line, which on closer examination was resolvable into minute dots or pustules, and the sensation was much as if each dot were charged with a red-hot needle gradually making its way through the nerves. The slightest touch of the clothes was agony, and as I had to walk more than two miles before reaching my lodgings, the sufferings endured may be better imagined than described.

Severe, however, as was this pain, it was the least part of the torture inflicted by these apparently insignificant weapons. Both the respiration and the action of the heart became affected, while at short intervals sharp pangs shot through the chest, as if a bullet had passed through heart and lungs, causing me to fall as if struck by a leaden missile. Then the pulsation of the heart would cease for a time that seemed an age, and then it would give six or seven leaps, as if it would force its way through the chest. Then the lungs would refuse to act, and I stood gasping in vain for breath, as if the arm of a garotter were round my neck. Then the sharp pang would shoot through my chest, and so *da capo.*

After a journey lasting, so far as my feelings went, about two years, I got to my lodgings, and instinctively sought for the salad-oil flask. As always happens under such circumstances it was empty, and I had to wait while another could be purchased. A copious friction with the oil had a sensible effect in alleviating the suffering, though when I happened to catch a glance at my own face in the mirror I hardly knew it —all white, wrinkled, and shrivelled, with cold perspiration standing in large drops over the surface.

How much brandy was administered to me I almost fear to mention, excepting to say that within half an hour I drank as much alcohol as would have intoxicated me over and over again, and yet was no more affected by it than if it had been so much fair water. Several days elapsed before I could walk with any degree of comfort, and for more than three months afterwards the shooting pang would occasionally dart through the chest.

Yet, as before mentioned, the result might have been more disastrous than was the case. Severe as were the effects of the poisoned filaments, their range was extremely limited, extending just above the knee of one leg, the greater part of the right arm, and a few lines on the face, where the water had been splashed by the curling waves. If the injuries had extended to the chest, or over the epigastrium, where so large a mass of nervous matter is collected, I doubt whether I should have been able to reach the shore, or, being there, whether I should have been able to ascend the cutting through the cliffs before the flowing tide had dashed its waves against the white rocks.

It may be easily imagined that so severe a lesson was not lost upon me, and that ever afterwards I looked out very carefully for the tawny mass of fibre and membrane that once had worked me such woe.

On one occasion, after just such a gale as had brought the unwelcome visitant to our shores, I was in a rowing boat with several companions, and came across two more specimens of *Cyanea capillata*, quietly floating along as if they were the most harmless beings that the ocean ever produced.

My dearly-bought experience was then serviceable to at least one of my companions, who was going to pick up the medusa as it drifted past us, and was only deterred by a threat of having his wrist damaged by a blow of the stroke oar.

Despite, however, all these precautions, I again fell a victim to the Cyanea in the very next season. After taking my usual half-mile swim I turned towards shore, and in due course of time arrived within a reasonable distance of soundings. As all swimmers are in the habit of doing on such occasions, I dropped my feet to feel for sand or rock, and at the same moment touched something soft, and experienced the well-known tingling sensation in the toes. Off I set to shore, and this time escaped with a tolerably sharp nettling about one foot and ankle, that rendered boots a torture, but had little further effect. Even this slight attack, however, brought back the spasmodic affection of the heart; and although nearly fourteen months have elapsed since the last time the medusa shook her venomed locks at me, the shooting pang now and then reminds me of my entanglement with her direful tresses.

For the comfort of intending sea-bathers, it may be remarked that although the effects of the Cyanea's trailing filaments were so terrible in the present instance, they might be greatly mitigated in those individuals who are blessed with a stouter epidermis and less sensitive nervous organisation than have fallen to the lot of the afflicted narrator. How different, for example, are the effects of a wasp or bee sting on different individuals, being borne with comparative impunity by one, while another is laid up for days by a similar injury. And it may perchance happen that whereas the contact of the Cyanea's trailing filaments may affect one person with almost unendurable pangs, another may be entangled within their folds with comparative impunity.

As, however, even the comparative degree is in this case to be avoided with the utmost care, I repeat the advice given in the earlier portion of this narrative, and earnestly counsel the reader to look out carefully for the stinger, and, above all things, *never to swim across its track*, no matter how distant the animal may be, for the creature can cast forth its envenomed filaments to an almost interminable length, and even when separated from the parent body, each filament, or each fragment thereof, will sting just as fiercely as if still attached to the creature whence it issued. It will be seen, therefore, that the safest plan will always be 'to keep well in front of any tawny mass that may be seen floating on the waves, and to allow at least a hundred yards before venturing to cross its course. Perhaps this advice may be thought overstrained by the inexperienced.

' Those jest at scars who never felt a wound,

but he who has purchased a painful knowledge at the cost of many wounds will deem his courage in no wise diminished if he does his best to keep out of the way of a foe who cares nothing for assaults, who may be cut into a thousand pieces without losing one jot of his offensive powers, and who never can be met on equal terms.

L

MY TOADS.

'THE toad,' observes an old and quaint writer, ' is
the most noble kind of frog, most venomous, and remarkable for courage and strength,' such qualities being
evidently indicative of nobility in the mind of the
narrator.   So among the Hindoos the cobra is honoured
as the creature of highest caste next to the Brahmin,
and an old and very vicious Hoonuman is deeply re-
spected as a very high caste monkey ; and so, throughout
all Oriental nations, the surest road to respect is to insult
their chiefs and thrash the people in general, giving no
reason for either proceeding.   In the present case, however, there are but little grounds for the respect that
our author evidently entertained for the toad, as, after
a long and somewhat intimate acquaintance with this
batrachian, I have found his venom impotent, have
never witnessed any display of his courage, and think
his strength to be, bulk for bulk, inferior to that of a
frog.   Still the toad is a respectable animal enough,
and to those who will wisely discard the prejudice
attached to its name, a very curious and interesting
animal.   Ever since I used to potter up and down our

garden, a small six-year-old naturalist, with a magnifying glass always open in one hand, and an empty pill-box in the other, the toad has been a great favourite with me, though, perhaps, considering the rude handling to which it was continually subjected, the feeling was hardly reciprocated on the part of the reptile.

En passant, let me speak in the highest terms of the benefit conferred on children by letting them run about as they will in a rough and ready kind of garden, where they may work their own sweet wills, dig, plant, sow, build, and play just as they like, without being subjected to the annoyance of being confined to the gravel, and forbidden under severe penalties to place a foot on the beds. It is an education in itself to them, this wild freedom. They learn a thousand things that books will never teach them; the use of their limbs, the use of their eyes, readiness of resource, and quick appreciation. They are sure to realise in vivid action every event of which they hear or read, and thus indelibly fix their knowledge on their childish memory.

For my own part I know that there was not an event in Robinson Crusoe, the Swiss Family Robinson, Persian Fables, or Arabian Nights, that we did not act over and over again; while the histories of England, Greece, and Rome were delineated with equal force.

Not only were we wrecked on desert islands—not only did we rescue Men Fridays (darkening our faces with black-lead, in order to represent the suave savage in ' character ')—not only did we build ' falcons' nests ' in the apple tree, and make rope-ladders out of clothes-lines—not only, in fine, did we reduce to practice any practicable event in our favourite books, and 'make believe' fervently in all impracticable cases, but we pursued the same system with severer studies, and acted in turn every historical person of whom we read, though the originals might have found some difficulty in recognising their representatives, or the localities in which the particular adventure occurred. For us, however, the result was perfectly satisfactory. If we pushed each other out of the stable-loft window, the Tarpeian rock was sufficiently indicated; and if the representative of the criminal happened to hurt himself by the fall it only made things look more real. And so, whether we gained our kingdoms by seeing flights of vultures, killed our brothers for jumping over the wall, got killed ourselves by an arrow in the eye at an imaginary Hastings, or one through the heart in an equally imaginary New Forest (the rocking-horse being of great service in the latter catastrophe), we certainly contrived to impress on our minds a tolerably vivid idea of the circumstances.

Children thus learn at the earliest years to distinguish one plant from another, to know a flower from a weed, and to learn something of their various properties; while, with regard to the animal kingdom, they gain a fund of practical experience that is sure to be valuable in after life.

It is no small matter for them to get rid of a fear,

to distinguish between the harmless and hurtful beings, and, by watching their interesting habits, to feel a sympathy with their fellow creatures, and to appreciate too keenly the infinite value of life to kill any living thing without just cause.

We were never afraid of black-beetles, daddy longlegs, or any of the insect tribe, except the few that wore stings ; while the frogs and toads were our special pets, lived in magnificent edifices made of bricks and flowerpots, and each had its own name.   Long before we read about them in books, we knew all about their absorption of water through the skin, their sudden cry of fear when alarmed, the equally sudden change of colour, and the curious fact that a frog which lived in a dark hole was always brown, and one that lived in the open air was yellow ; while as to the venomous nature of the toads, as energetically detailed by our nursery maids, we treated the notion with supreme contempt, and handled a toad as unconcerned as if it had been a ball. I am sure that many persons—young ladies especially—who cannot rid themselves of real terrors at the sight of many a harmless and useful creature, would have been much happier if their early lives had been spent in a garden such as has been described.

Having always felt an interest in these ungainly but truly useful batrachians, I begged from a friend a fine pair of Natterjack toads that had just been sent from Jersey, and placed them in a glass fern-case.

Their first proceeding was to establish hiding-places,

each choosing its own corner for that purpose. The method in which a toad ensconces itself is rather curious. Supposing, for example, that it wishes to burrow into the base of a small mound, it begins by finding some small part where the earth is tolerably loose. It then plants the extremity of the back against the mound, wriggles about in a position that reminds the observer of the green crab shovelling itself under the sand, and pushes the earth from beneath it with the hind feet, passing it forwards under the body, where it is taken up by the fore-feet and put out of the way. Inadequate as the means may seem—the soft, skinny feet of the toad being apparently the worst spades that could well be devised—the creature will sink itself below the ground in a wonderfully short space of time. It is remarkable that a toad never enters its hole except by backing into it—at least, I have never seen one do so, whether it be at liberty or in confinement.

Having fairly established themselves, they looked out for food, although, with all of their kin, they were capable of sustaining a very prolonged fast without any apparent inconvenience. As at that time I was living in the very heart of London, it was not easy to procure the proper kind of food for the toads, who feed wholly upon living creatures, and will touch nothing that does not move. However, I contrived to bring home a miscellaneous collection in several boxes, and tried experiments with them.

They would eat earthworms, provided that they

were clean and lively, so as to writhe about in that manner which a toad cannot resist. They were captured after the usual custom—namely, by a sudden 'flick' of the curious tongue, which is so rapidly moved that with the most careful attention the eye can only distinguish a pink streak appear and as suddenly vanish. A slight slapping sound is heard as the tongue is thrown towards the prey.

The fact has long been known, but the details have, I believe, not yet been fully described. It is by no means necessary, as has been repeatedly asserted, for the toad to remain motionless, with its eyes intently fixed on its victim. On the contrary, I have often seen the toad catch beetles in spots where it could not see them, and without even attempting to look for them. The tongue can be flung in any direction, and always with equal certainty of aim, at right angles to the head for example, out of either corner of the mouth, or even under the body. I have repeatedly seen the creature aim at an insect that was crawling under its body, and mostly with success; if not so, a second shot was sure to be effectual.

I used frequently to feed them with blue-bottle flies, by the simple process of putting them into the fern-case and closing the entrance. In spite of the wings and activity of the insect, the toad was sure to have it before long. At the first buzz the toad would come all in a hurry out of his hole, tumbling over stones and sticks in his eagerness, and evidently listening for the sound of the fly's wings. As soon as the insect settled within reach of the tongue (and when the reptile stood on its hind legs it had a marvellous reach) the toad used to raise its head with an oddly knowing air, and looked as eager as a cat which hears a mouse behind the door. It would then scramble hastily towards the fly, when a red streak would be seen to flash from its mouth, a slight slap was heard, and the fly had vanished. If the insect took alarm, the toad was quite content to wait, and was certain to hunt it down at last.

It may be here mentioned that the root of the toad's tongue is set on the front of the lower jaw, the point being directed backward, so that when an insect is captured the mere return of the tongue flings it down the throat. A few decided gulps are, however, needful to complete the operation, and the aspect of the toad while engaged in swallowing is most absurd, the elevated eyes being closed, and disappearing entirely by the exertion. The dimensions of the insect make no difference in the magnitude of the gulp and the disappearance of the eyes.

Few persons who have not watched a toad can form any idea of the dexterous manner in which it uses its fore-paws, these apparently clumsy members serving the purpose of hands, and being frequently employed in lieu of those important limbs. If, for example, the toad has snapped up a tolerably long worm, it will probably be incommoded by the natural objection

entertained by the annelid with respect to its lodge-
ment in its captor's stomach, and the struggle which it
makes to escape, its head and tail usually protruding
at opposite sides of the mouth.

Now the toad is strangely indifferent to wounds
and injuries, and even if nearly severed in two seems to
be as unconcerned as if it had no personal interest in the
calamity. But nothing appears to annoy the strange
creature so much as any object sticking in the sides of
the mouth, and it displays a vast amount of uneasiness
until it has removed the annoyance. In order to effect
this object the fore-paws are brought into play, the
creature grasping at the irritating object just as a
monkey would do under the circumstances, and either
pushing it down the throat or throwing it away, according to its fitness or unfitness for food. I have known
the leg of a beetle, or even the wing of a fly, worry the
toad sadly, while a small blade of grass excited it to
such a degree that it very nearly looked angry.

There is one curious point connected with the toad
which I have never been able to comprehend. Supposing it to be pursuing a fly, and the insect to have
settled out of reach, the toad sits watching it just as a
lion is said to watch a baboon or a human being who
takes refuge in a tree. While thus watching, the last
joint of the middle toe of the hind feet is continually
jerked with a convulsive kind of movement, twitching
in unison at regular intervals. The movement seems
to be quite involuntary, and I suppose is analogous to

the waving of the lion's tail while the animal is crouching in view of its intended prey.

Although the toad can endure a very long fast, there seems to be no limit to its gormandising capacities when it meets with a plentiful supply of food. The smaller of my specimens ate successively several worms, a great ' woolly bear ' caterpillar (*i.e.*, the larva of the tiger moth, *Arctia caja*), a large grub, apparently the larval state of some beetle, a number of smaller insects, and a large ground beetle (*Carabus violaceus*). These various capabilities render it a most useful animal, and one which should be carefully guarded by every owner of a garden. For at night, when the obnoxious slugs, flies, beetles, and other insects are on the move, the toad comes out to prey on them, and quietly performs very great service by the steady, thorough-going manner in which it clears the plants of every creature that moves.

Some entomologists, whose zeal for the enrichment of their cabinets exceeds their humanity, are in the habit of sallying out into the fields at early dawn, killing all the toads that they can find, and opening them for the purpose of getting the insects that have been swallowed during the night. Some of the rarest British specimens have been taken in this manner, beetles being the usual denizens of the locality. Conchologists are accustomed to employ a similar mode of collecting the objects of their research, and find some of the best specimens in the stomachs of several deep sea fishes ; and microscopists in like manner find a vast museum of beautiful objects within the digestive organs of various molluscs.

The beautiful eye of the toad is proverbial, redeeming the ungainliness of its general aspect, and having in all probability given rise to the fabled jewel within the head. Bright and richly coloured as is the eye, with its glowing, bold, fiery chestnut hue, it is without the least vestige of expression, and retains its full brilliancy long after the animal is dead. As to the venomous powers of the toad, they are not to be found in the mouth, as is popularly imagined, but in two rather large glands on the sides of the head, which project boldly, and are plainly visible. If one of these protuberances be squeezed between the fingers, a whitish, creamy-looking liquid will be ejected, and perhaps to some little distance. While performing this operation it will be as well to hold the toad in such a manner that the secretion may not be shot into the eyes, as in that case it would probably cause severe pain, and might probably produce violent inflammation. Still it will not be ejected without the employment of considerable force, and is never injurious to human beings.

Briefly to sum up the character of the toad: it is not pretty, is entirely harmless, extremely useful, easily tamed, and worthy of being cherished by those who prefer deeds to outward seeming; it is a creature of curious and interesting habits, and affords a rich field to anyone with time and opportunity for clearing up several important but disputed points in physiology.

THE CHILDREN OF THE NEW FOREST.

IT was a delicious summer evening, the fresh breeze
pouring new life into lungs choked with thick London
smoke, and the setting sun darting its last red rays
through the waving corn, when we issued from the
station door, wearied and cramped with long sitting in
a crowded carriage, and were heartily greeted by our
host, whose domains we were about to invade.   A few
minutes served to settle us on the vehicle in waiting,
and the train had hardly proceeded on its course when
we were merrily bowling along towards our home in the
New Forest.

Even the country drive was a luxury to those who
had for months been penned up in the very heart of the
metropolis, and, after a mile or so had been passed,
proved to be not without its excitement. The favourite
old horse—hight Rufus, in honour of the second William,
and in allusion to his bay coat—trotted off in great
spirits, knowing that every step took him nearer to his
stable. His owner, however, not wishing us to be taken
by surprise, mentioned casually that Rufus generally
fell down when descending a hill, and that he always

liked to have the vehicle pushed behind him whenever he came to an ascent; but that those who were used to him knew what to expect, and did not object to these trifling eccentricities. As, therefore, the road consisted, on an average, of six miles of hills and two of level ground, it may be imagined that mental excitement was combined with physical exertion in a degree rarely witnessed. However, we had started with the intention of taking everything as it came, and therefore held up Rufus carefully as he went down hill, and pushed behind when he went up hill, until we arrived at our intended domicile; the vehicle having been very useful in holding our baggage, but as far as ourselves were concerned, rather an honourable appendage than a personal conveyance.

Evening had set in long before, and the glowworms had started one by one into their full beauty, as they lined the forest pathways like mundane stars shining in imitative rivalry of the glittering points in the dark dome above. One of them we placed on the splashboard by way of a lantern, and on our safe arrival laid it carefully among some herbage just outside the door, a position which it held for three days and nights.

Such a lovely spot is the New Forest; the soil so various, the trees so magnificent, the flowers so perfumed and luxuriant, and the birds so plentiful and musical. May the Enclosure Act, that has turned many a mile of grand forest into base turnip land, never lay its wither-'ng grasp on the New Forest! and far be from our eyes the chilling sight of the splendid oaks, which we have so long loved, lying like murdered corpses on the ground, their white and gnarled limbs stretched out as if stiffened in deadly agony, and their rugged bark, erst rich with moss and lichen, stacked in heaps by their sides.

Some unimaginative persons talk of the dull uniformity of the forest—you might as well talk of the dull uniformity of the Strand or Regent Street, and with much more reason of the dull uniformity of Rotten Row. The real deep, primitive forest is ever changing, and in one day may pass through a thousand phases. Putting aside the two great epochs of summer and winter, of leafless branch and wealthy foliage, of green-clad boughs and snowy shroud, together with the intermediate state of spring's delicate green and autumn's ruddy brown, there is hardly a day when the forest does not assume a new aspect as each hour passes away, and in which its threefold harmonies of sight, sound, and scent are not woven into a thousand varied modulations, like a fugued melody of some great master in music. Mendelssohn always reminds me of a forest. No one can appreciate a forest who has not passed whole days in its solitary depths, and watched it from the early morning hours to the deep, dark shades of night. Different birds, insects, and flowers make their appearance at their chosen hour, and there are many creatures which emerge from their hiding-places only for a brief space, and then return into darkness and solitude for the remainder of the day. The sweet voices of the song-birds have their appointed times, and the perfume of flower and leaf changes with the march of the sun.

Full of pleasant memories, and gay with anticipations of the morrow, we two old foresters flung open our window to the utmost, so as to be lulled to sleep by the owl and the silence, and to be awakened by the merry songs of the morning birds. We awoke at the intended hour, but heard no birds, nothing but a rushing sound as of rain on leaves. Horror! the sky is of one uniform leaden tinge, and the rain is pouring in steady perpendicular torrents, as if a second deluge were impending. What shall we do for the next few hours, while the household is asleep within and the rain pouring without? Let us brave the storm, accept a thorough soaking as an inevitable fact, and sally boldly into the forest just to see its aspect after a wet night and during a heavy rain.

A few minutes served to encase ourselves in the very oldest habiliments that our wardrobe could furnish, and to see us on our way. Twenty yards sufficed to drench our clothing as effectually as if we had just emerged from the depths of a river, and from that moment we became delightfully indifferent to the rain; having a kind of wild exultation in the feeling that we could walk about in the midst of the watery torrent without seeking shelter or needing an umbrella. I have seldom enjoyed a walk more than that saunter in the forest glades, with the noisy patter of the rain-drops on the leaves overhead, the pleasant smell of the crushed fern, the primitive independence of being thoroughly wet and caring nothing for it, and the plish-plash of our feet as every step pumped water out of our boots. Back to the house through the rude path, now some six inches deep in red mud, a brief toilet, and a very welcome breakfast.

Still rain, rain, more rain, and what shall we do? Cat's cradle afforded a little amusement, uniting the advantages of adventurous combination, unexpected results, and the least possible bodily exertion. Even this recreation, however, is scarcely exciting enough to be continued for any lengthened period, and after a desperate but abortive attempt to play at fives in an empty garret, we extemporised a game at bowls on the floor, the 'jack' being represented by a bradawl stuck in the boards, and the bowls by two india-rubber balls, one solid and small, and the other hollow and large.

The beauty of the game was enhanced by the sloping nature of the floor causing the balls to roll away until they were either checked by the wall or fell down the staircase. This difficulty, however, was overcome by the inventive genius of one player whom modesty forbids me to particularise, and a few handfuls of oats scattered over the floor served at once to arrest the ball and to test the player's skill in guiding his bowl neatly into the little hollows left here and there by the grain. This absorbing pursuit carried us over three or four hours, when its course was suddenly arrested by a summons to dinner, the greater part of that refection having been cooked in the solitary sitting-room of the establishment.

Rain still heavy, if anything heavier than before, and what shall we do? Let us throw knives at a mark like Ho Fi, the Chinese juggler, whose portrait we had lately seen, represented as in the act of aiming a broad-bladed knife at a fellow-countryman standing spread-eagle-wise against a board, and whose outstretched limbs and rigid head were encircled with similar weapons. No sooner said than done. A target was rapidly improvised, a stout board fetched from the shed, a couple of 'rymers' sharpened, and in a few minutes all hands were deep in this most absorbing pursuit, which, when afterwards imported into the metropolis, proved of so fascinating a character that I have known the whole male population of a drawing-room desert their fair companions, and give themselves up an unresisting prey to 'pegging.' Nothing is simpler than this game. You take a sharp-pointed knife, chisel, or other implement, lay it flat along the hand, the point directed up the arm, and the handle just projecting from the finger-tips. You take a good aim at the target, fling the knife so as to cause it to make one half-turn as it passes through the air, and if you have performed all these actions correctly, the knife darts into the target with a heavy thud, and there sticks, quivering with the violence of the blow. It is, in fact, a refinement on 'Aunt Sally;' quite as exciting and not half so fatiguing.

Night again drops her dank, wet veil over the scene, and our visit to the New Forest bids fair to be a total failure.

Brightly shone the sunbeams on the following day; the dismal splash of rain had ceased; the black, cloudy sky had changed to deep blue; the breeze was charged with perfume, and the air filled with melody. A host of chaffinches were congregated in front of the window, pecking about amongst the grass, and twittering merrily with their sweet little chatter. All nature seemed to rejoice in the sunshine, and the deep glades of the forest, broken by sundry gleams of golden light, invited us to its presence.

The ground was still wet under our feet, the heavy ferns dropped showers of moisture as we brushed against their wide fronds; and, as the wind stirred the branches above, occasional shower baths came pattering on our heads. But how changed was everything around. The birds flitted from bush to bush, heedless of the rain-drops scattered by their rapid movements; the air was filled with glittering insects, and the busy hum of many wings gave light and brightness to the scene. The long avenues of oak and beech produced effects of brilliant many-coloured light and deepest shade that no painter could hope to imitate; the heavy masses of holly that studded the forest gave a mysterious darkness to many an inlet, while the wide clusters of foxgloves reared their tall heads in the patches of sunshine, and waved their lovely petals in the breeze. Foxgloves, indeed, seem to be the leading characteristic of that part of the forest, for it was impossible to look down any avenue without seeing a cluster of these magnificent flowers shining out against the dark masses of shadowy verdure, and giving wondrous effects of colour just where an artist would most want them

It was most beautiful, too, to watch the golden-winged insects come darting against the sunbeams, issuing like visions from the shrouded darkness, glittering for a moment like living gems as they shot through the narrow belt of light, and vanishing into the mysterious gloom beyond, as if suddenly annihilated by the wave of a magician's wand. More pleasant to the sight than to the touch, particularly for persons endowed with a delicate skin. I never thoroughly appreciated the exceeding torture that the plague of flies must have inflicted upon the Egyptians until I had passed a few hot summer days in the New Forest.

Flies of all sorts, sizes, and colours surround the hapless victim, and render existence a burden and a torment. Great buzzing, wide-winged, large-eyed flies charge at him with a trumpet of defiance, and in spite of clothes find some weak point through which they may insert their poisoned dart. Tiny flies, too small for audible murmur of wings, and too gentle of move-ment to be noticed, run nimbly about his person, creep up his sleeves, slip down his neck, get into his eyes and nostrils, and leave memorials of their presence in a series of little angry red pustules like those of nettle-rash, and quite as annoying. Others, again, will set to work in a calmly composed and business-like style, alight on his hand or wrist, produce a case of lancets from their mouths, and bleed him with the practised skill of an old surgeon.

Besides all these foes, the forest is haunted by myriads of horrid ticks—flat-bodied active little creatures, with legs that cling like burrs, and heads barbed like the point of a harpoon. These insidious animals swarm upon the passenger, and are sure to discover some method by which they may creep through the clothes and operate on their victim. Imperceptibly the barbed head is thrust under the skin, and the creature begins to suck the blood of its human prey with such voracity that before many minutes have passed, its flat and almost invisible body swells into a blood-distended bag, and the tick looks more like a ripe black currant than an insect. If it should be discovered it must in no wise be torn away by violence, or its barbed head would remain in the wound and be the cause of painful inflammation.

There are two modes of ridding oneself of ticks. One method is by lighting a large fire, taking off all clothing, and rotating before the blaze as if attempting suicide by roasting. The ticks cannot endure the heat, and soon fall off, but as this process is hardly feasible in an English forest, it is better to have recourse to the second method, which is simply to brush them with a feather dipped in oil.

As for myself, in spite of wearing large gauntleted leather gloves, and tying the wrists and ankles with string, the insects led me such a life that I hardly dared enter the forest. At last a bright idea struck me. I rubbed my hands, ankles, face, and neck with naphtha, and kept a little bottle in my pocket for renewal whenever the odour seemed to become faint and ineffectual. After taking this precaution I enjoyed a delightful immunity from insects, which more than compensated for the very unpleasant smell of the naphtha. Even in the course of a long day's sojourn in the forest depths not a fly dared meddle with so potent an odour, and it was amusing to see a great loud-winged insect come charging along, ready for action and thirsting for blood, and then to see it pause in full career, balance itself for a moment on quivering wings, and dart off at an angle from the hateful scent.

Upon many a tree were the nests or ' cages ' of the squirrel, denoting the abundance of these pretty little animals in the neighbourhood. Before very long a reddish dot was seen moving among the grass, and we immediately determined to stalk up to the creature and to watch its habits. Being accustomed to woodcraft, and knowing how to take advantage of every cover, to pass among branches without noise, and to avoid snapping dried sticks with the feet, we crept to a tree-trunk within six yards of the squirrel, and there sat quietly looking at him.

There he was, blithe and joyous, totally ignorant of

our presence, but still watchful, raising himself occasionally so as to look over the tops of the grass-blades, but never seeing us on account of our rigid stillness. It was most interesting to watch the pretty little animal as he went skipping over the ground in little hopping steps, now stooping to feed, now picking up something in his paws, holding it to his mouth in a dainty and well-bred fashion, tasting it, and then throwing it down in disdain. Then he would disappear entirely below the grass, and next moment he would be sitting upright, his bushy tail curled over his head, and his bright eyes gleaming as he looked around.

Suddenly a lad came running towards us, making much more noise in crashing through the fern than a dozen full-grown elephants would have produced. Up jumped the squirrel, glanced hastily towards the spot whence the unwelcome sounds proceeded, and dashed off for the nearest tree, looking wonderfully like a miniature fox as he scudded over the ground, his body stretched to its full length, and his bushy tail trailing behind him. A long leap, and he had jumped on the trunk of the tree towards which he was running, and, according to the usual fashion of squirrels, skipped round it, so as to interpose the trunk between himself and the supposed foe. But this manœuvre exactly brought him face to face with us, and at the distance of only a yard or two, and I never saw a squirrel look more bewildered than he appeared on making this terrible discovery. He never stopped for a moment, however, but fairly galloped up the tree, ran along a projecting branch, made a great leap into another tree, traversed that also, and in two minutes was fairly out of sight.

Here let me offer an indignant protest against two subterfuges under which the destroying nature of man hides its ugliness.

There are some persons in whom the destructive element is acknowledgedly developed in all its fulness, who live but to hunt, and to shoot, and to fish, and who really seem to have gradually drilled themselves into a heartfelt belief that to destroy the furred, feathered, and scaled inhabitants of the earth is the noblest aim of man, and one to which every other object must necessarily become subservient. As a natural corollary of this proposition, follows the extirpation of every living creature that can interfere, either actively or passively, with their sport, the result being to depopulate the country of every being in which is the breath of life.

All the beautiful, and truly beautiful, weasel tribe are to be killed because they will eat hares, rabbits, and feathered game; all the hawk tribe fall under the ban; the ravens, crows, and magpies are to be killed because they are apt to rob the nests of partridges and pheasants; the little birds because they eat the corn on which the pheasants might feed; and even the squirrel is now reckoned amongst the vermin because it is known to regale itself occasionally on young birds, and possibly on their eggs. The keeper who destroys the greatest number of these ' vermin ' earns the highest praise from his master; and, to all appearances, the very perfection of a forest in the eyes of a sportsman would be this—it should not harbour a single creature except those which are dignified by the title of game, and thought worthy of death from the hand of their owner.

It is a pitiful sight in this grand forest to view the handiwork of the keepers in the shape of noble hawks, ravens, martens, squirrels, and other wild denizens of the woods, nailed on the trunks of trees, or hung in withered clusters from their boughs. I do not believe that a true sportsman would find his amusement curtailed by their life, feeling sure that nature can generally keep her own balance, as is exemplified in countries where the Game Laws were never heard of, where game preservation has never been dreamed of, and where the game abounds in spite of the swarming vermin far more numerous and powerful than those of our own country.

Another and more noxious kind of destroyer is found in those pseudo-zoologists who hypocritically conceal their love of slaughter under the guise of science, and, necrologists as they are, never can watch an animal without wanting to kill it. The daily papers afford abundant instances of such mock science, and it is well known that even a parrot cannot escape from its domicile without running the most imminent risk of being shot. Not a rare bird has a chance of escape if it once shows itself within the limit of the British Isles; and I can but think with exultation of those deluded individuals who lately spent much powder and shot, and more patience, upon some rare sea-bird which had settled in a lake, and which afterwards proved to be nothing but a stuffed skin ingeniously anchored by a long line. Such persons never think of watching the living being in order to learn the wonderful instincts with which its Maker has gifted it, and the interesting habits and customs belonging to the individual or the species. Should they come across a rare bird, their first regret is that they have no gun with them; and instead of feeling delighted at the opportunity of gaining further knowledge, they only lament that they cannot take away from the bright being that life which it is so evidently created to enjoy, and the causeless deprivation of which is literally a robbery of its birthright.

One of the principal objects of our expedition was to ascertain the mode in which the snipe produces the remarkable sound called ' drumming,' from its fancied resemblance to the distant roll of the military drum. To my ears, however, the mingled whizz and hum of a slackened harp-string give the best idea of this remarkable sound.

It must be premised that during the breeding season the male snipe, like many other creatures, assumes new habits, and utters new sounds. Generally the flight of this bird is short and fitful, as is well known to all sportsmen, and seldom lasts more than a few minutes. But during the breeding season the snipe becomes an altered being. Towards evening it leaves its marshy couch, and rises to a great height in the air, where it continues to wheel in circuitous flight for a considerable period, mostly confining itself within the limits of a large circle, and uttering almost continually a loud, sharp, unmusical, and monosyllabic cry, which may be roughly imitated by the words 'chic! chic! chick-a chick-a, chic! chic!' &c. At varying intervals it sweeps downwards, making a stoop not unlike that of a hawk, and producing the sound called drumming during the stoop.

How the bird drums has long been a matter of doubt, some naturalists attributing it to the organs of voice, others to the wings, and others to the tail. To set this question at rest was therefore an interesting pursuit, and to that purpose several successive evenings were devoted.

As soon as the snipes began to drum we set out for the marshy ground over which they flew, and by dint of cautious management succeeded in ensconcing ourselves in a dense thicket of thorn and blackberry where we were perfectly concealed, but whence we had a thorough command of the sky. Not choosing to trust to my single observation, I had two friends with me; one of them a well-known bush huntsman of Africa, and the other an old and observant inhabitant of the forest. We were all supplied with powerful glasses.

Before we had lain very long in ambush the desired sound struck our ears, proceeding from a snipe that was circling high above us. We watched the bird for a long time, but he never came near enough to give a good view. Several others afforded us much disappointment, but at last all our trouble was fully repaid. A fine snipe arose at no great distance, and just as if he had known our object, and intended to give us his best aid, began to cry and drum just over our heads, and at so small a height that as he wheeled in airy circles his long beak and bright eye were clearly seen even by the unaided vision, while the double field-glasses with which we were supplied gave us as excellent a view of the bird as if it were within two yards.

It was, then, quite clear that the drumming sound was not produced by the voice, as the bird repeatedly uttered the sound of ' chic ! chic ! chick-a ! ' simultaneously with the drumming. Without offering any opinion we repeatedly watched the bird, and then compared our observations. They were unanimous, and to the effect that the sound was produced by the quill feathers of the wings. The bird never drummed except when on the stoop, and whenever it performed this manœuvre the quill feathers of the wings were always expanded to their utmost width, so that the light could be seen between them, and quivered with a rapid tremulous motion that quite blurred their outlines. Our observations were repeated during several successive evenings, and always with the same result.

There is perhaps no locality in the whole of this country so well adapted to the natural historian as the New Forest, the conditions of soil, elevation, and foliage being so prodigally varied that almost any creature can find a refuge in some portion of its limits. Take, for example, the spot on which we resided, but which I do not intend to particularise, lest its sacred recesses should be profaned by the step of outer barbarians, and its wild glades polluted by empty porter bottles, broken crockery, and greasy sandwich papers.

The cultivated ground in front of the house reached a narrow and rapid brook. Beyond the brook was a large expanse of marsh and shaking bog, harbouring multitudes of snipes. In the middle of this swamp our drumming observations were made. The ground suddenly rose beyond this bog into a wide but not very high hill, covered densely with heather, and giving shelter to grouse and pheasants. About four miles further the heath was abruptly ended by a large fir-wood, in which the deer loved to couch. We once devoted a whole morning to tracking a deer by its footsteps or 'spoor,' and after some three hours' careful chase found the creature lying couched among the fern. Ravens were often seen heavily flapping their way over the heather, and on one or two occasions our eyes were gratified with the grand sweeping flight of the buzzard, as it soared on steady wing, inclining from side to side like an accomplished skater on the outside edge, but appearing to make its way through the air as if by simple volition. Bright-plumaged woodpeckers fled screaming through the forest depths, and many a tree-trunk bore witness of their persevering labours.

The human population of the forest have, in course of time, become deeply saturated with the wild uncultivated air of the region in which they reside, and many an aged man has never seen a town in his life, or ventured beyond the limits of the familiar forest lands. A practised eye can mostly detect a forester at a glance, a strange family likeness being observable in all who have passed their existence in this place—probably owing to the continual intermarriages which necessarily occur among them. Even the tone of voice is of a peculiar nature, and the drawling, high-pitched chant of the thorough-bred forester is not likely to be forgotten if once heard. In fact, the forest is to its aborigines what the desert is to its nomad Arabs; and the wild Bedouin can hardly feel more terror at the idea of entering the habitations of civilised man than does the forester at the notion of exchanging the trees for houses.

I remember that on one occasion, after the hay had been got in, a cartload was destined for some stables at Southampton. The fragrant trusses were placed on the waggon, the horses harnessed, and all was ready for the journey, when an unexpected difficulty arose in the person of the carter, a fine young fellow of six-and-twenty, one of the first in the field and all the rustic sports. After a vast amount of prevarication he flatly refused to leave the forest, and when peremptorily ordered to do so he sat down on the road-side, and sobbed like a child with sheer terror of the unknown regions beyond his ken. An exact parallel to his despairing fears may be often seen in the crowded thoroughfares of London, where a child has lost its way, and stands weeping in the depths of its misery, beset on all sides by vague fears, and as hopelessly bewildered as if it had been suddenly transplanted to a new planet. Take such a man out of the forest, run him off by express train to London, put him down at London Bridge or Charing Cross, and he would become a maniac from the rush of ideas to the brain, like that Kaffir chief whose head was turned by the sights on board of a steamer, and who deliberately hauled himself to the bottom of the sea by means of the chain cable.

There is also a strange race of beings called the woodmen, who possess certain prescriptive rights from time immemorial. They are the most independent set imaginable, and laugh at law or justice. Their carts are at least two feet wider than is allowed by legal authority, and while driving along the road they are totally regardless of the right and wrong side. Those who meet them may turn aside if they like, but they proceed on their course without paying the least respect to the tacit regulations of the road. One Saturday, while driving on the high road, we met a long string of woodcarts all on the wrong side, all straggling in such a manner that we were fain to draw our vehicle into the ditch, and on every cart were one or two wood-men lying in a state of senseless intoxication, and leaving their horses to find their own way home—a task which they certainly performed with an accuracy that warranted the confidence reposed in them.

Many of these men would not be sober until the Tuesday, they would sleep off their headaches on Wednesday, on Thursday and Friday they would earn a week's wages, and on Saturday they would set off to the public-house and renew the last week's scenes. This kind of life suits their lawless natures, and they would rather lead this wild and reckless existence than become honoured and useful members of society, as they might easily do, considering the wages which they can earn. Perhaps their wives and children might hold a different opinion, especially from Saturday evening to Wednesday morning.

Vipers are delightfully plentiful in the New Forest, and during our limited sojourn I saw three distinct varieties, the common, the light grey, and the yellow, the last-mentioned being the largest living viper I ever saw. *Apropos* of vipers, it so happened that some farmers were paying a passing call when a labourer brought me a moderate-sized viper suspended to a string, and hung it on a post. Acting on the impulse of the moment I flung my knife at the reptile, and by a wondrously fortunate shot drove the blade fairly through the spine, just behind the head. My friend followed suit, and transfixed the snake about the middle of its body. The farmers were quite aghast at our skill, and it may be imagined we did not disabuse them of their good opinion by attempting a repetition of the feat.

After a number of experiments on the living viper I found that the reptiles could never be induced to bite at a stick, however great the provocation might be, but that as soon as any living creature came within reach they were sure to strike. The foresters were actuated by a wholesome dread of the viper, but feared the harmless blind worm far more than the really venomous reptile. One of the labourers brought to me the upper half of a blind worm squeezed tightly in his cap (the creature having thrown off its tail according to custom), and was almost pale with horror when I took it from the cap with bare hands. Mr. Waterton's feat of carrying twenty-seven living rattlesnakes from one room to another afforded a sufficiently terrifying spectacle, but in the eye of a genuine forester could not compare with the prowess displayed in seizing a blind worm with the bare hands.

Perhaps the night walks in the forest afforded the most pleasant reminiscences of our visit. At nightfall we used to put a compass and some matches in our pockets, and start for the depths of the forest, taking care to step very gently so as to give no audible alarm, and to keep ourselves well in the shade so as to avoid detection by sight. It was most delightful to wander thus into the heart of the primeval forest, among the great oaks and beeches, to seat ourselves face to face on the soft moss at the foot of some tree, and listen to the weird-like sounds alternating with the solemn stillness of the woods. At times the silence became almost audible, so profound was the hushed calm of night; while at intervals the sharp yapping bark of a fox might be heard in the distance, the drowsy hum of the watchman beetle came vaguely through the air, and the locust-like cry of the goat-sucker resounded from the trees. These curious birds were very common and quite familiar, allowing us to approach within twenty yards of the branch on which they crouched, or sometimes sweeping with their noiseless flight to the ground in front of us, and then pecking merrily away at the various insects which traversed the grass. There is, by the way, a curious superstition about these birds. If they come close to a house and sing three times, they prophesy a death in the family; if five times, a birth, and if seven times a wedding. It is strange that man and animals should fall so readily into the primitive life, and allow the instincts to regain their original and legitimate sway over the habits. Even the very cows learn the customs of the bush in a marvellously short time, and walk with the same lifted step as the ant lope that has spent all its life in the forest. One night, as we were concealed under the shade of a tree, a light crackling of dry sticks was heard. We drew deeper into the shadows, assured ourselves that nothing white was visible in our dress, and that our sticks were well grasped, for a night walk in the New Forest is not without its perils, the poachers being perhaps the most crafty and desperate set in England. Man or beast, however, the creature passed by, but kept so closely in the shade that we could not even catch a glimpse of its form. Stealing gently to the spot we felt the ground carefully, and soon found the fresh spoor of a cow, which had got into the forest, and instinctively moved as if it were liable to be hunted as soon as seen.

After a number of experiments we found that nothing is so utterly invisible in a forest at night as darkish grey, but not too dark. Black is seen with comparative ease, red is nearly invisible, and so is brown, but with dark grey the only visible portions are the hands and face, so that a pair of dark gloves and a dark mask would render a human being quite undistinguishable at two yards, provided he remained in the shade, and did not allow his form to be defined against the sky.

One night was truly memorable. We had started as usual, when we saw an odd kind of light among the trees for which we could not account. First we thought it was a paper lamp hung up by way of a trick, but soon found that it was far beyond the trees. Surely it must be Capella shining dimly through a fog, but on looking more carefully Capella was discovered without any fogginess about it. Suddenly my companion gave a hideous shriek, executed a *pas de seul* expressive of astonishment, and employing, as is his custom when excited, language more remarkable for energy than for elegance, cried out 'that it was a thundering big comet, as safe as the bank!' And so it proved to be. No more forest for us that night; but out came the telescope, the sextants, and the note-book, and the whole evening was passed in taking observations and running into the house to record them. As the comet stretched its mighty train over the zenith, great was the excitement as its vast proportions expanded with the darkness of night. 'I'll get its angles with the stars,' cried my friend, 'and you measure its length.' Off to the house at full run.

'How many degrees?'

'Eighty-two and a half.'

'Humbug! I don't believe it.'

'Look for yourself then.'

'Must have been wrongly handled; I'll measure it myself.'

Off rushes the excited astronomer, sextant in hand, and in five minutes is back again.

'How many degrees is it now?'

'Eighty-six I make it.'

And in this manner we spent the greater part of that night, the comet seeming to lengthen with every hour. It was certainly a most startling occurrence. No one expects to walk out of a house according to usual custom, and to meet a full-blown comet in the face. But here was the stranger, waving its flaming sword over our heads, and stretching its vast length over a greater space of sky than was occupied by the great comet of 1858, which had spent so many weeks in attaining its full size.

Much more is there to say of the New Forest, of the many-tinted flowers, its wealth of insect life, its wild and piquantly-flavoured fruits, and its wonderful depth of foliage, its grand old trees, among which the 'king beech' raises its royal head in token of superiority. It is indeed almost a new world, and to a Londoner affords a fresh current of ideas that regenerates the mind like fresh blood to the heart. Here all conventionalities cease. Mrs. Grundy could not live for five minutes in the forest depths; there are no neighbours to criticise the appearance, no gossips to decry the character. Man lives for a while the real unsophisticated life of Nature, and, it may be, will learn many a lesson for which he will be the better until his dying day, and perhaps after it. And those privileges may be gained by just taking a railway ticket for the nearest station to the Forest (say Southampton), where the traveller will be deposited in less time than is often occupied in getting to an awkward suburb of London. But our space is at an end, and we must reluctantly bid a farewell to that valued spot, hoping soon to visit it again.

A BLACKBERRY BUSH IN AUTUMN.

SOME years ago an Australian settler was on a journey, and, as he travelled, he saw what he took for the moment to be a blackberry bush. Of course it was not a blackberry bush, but the very semblance of the familiar bramble, with its well-remembered berries, so stirred his recollections of childhood, that he could not rest until he was on his way back to the old country. I can fully sympathise with his feelings, for I confess to a very strong affection for the blackberry, which I always visit whenever there is an opportunity, though I care not for its fruit.

Within a few yards of my house there is one blackberry bush for which I have the strongest admiration, and there are few days when I do not visit it. It stands by itself in a pasture field, of which it forms one of the most conspicuous features.

There are plenty of trees in the field. Nearest to the blackberry is a clump of horse-chestnuts, one of which is rapidly dying, having evidently been attacked by the great white grub of the stag-beetle, an insect which absolutely swarms in this part of the country, and which, as it passes several years in the larval state, and feeds all the while upon the tree, makes no small damage in the neighbourhood.

It is easy to detect the trees that are afflicted with this destroyer. They first show signs of decay in the upper branches, and the tree gradually dies down to the root. If it be cut down, and the stump opened, the cause of the disease will at once be evident in the shape of large soft, grey, white, shining grubs, with very large and lumpy bodies, and a pair of horny, powerful jaws, between which anyone who gets a finger will repent it, and so obese that like Basil Hall's fat pig 'Jean,' the creatures cannot stand on their little legs, but are obliged to lie perpetually on their sides.

At some distance is a row of silver birches, their shining stems glistening against the background of brackens which stretch beyond them; a tall, weeping birch waves its feathery plumage in the breeze, and around is a fringe of elms, oaks, and poplars, with one or two fine cedars spreading their 'dark layers of shade' in the middle distance.

It might seem that a simple blackberry bush would look quite insignificant in the midst of such surroundings; but, in reality, they only serve to set forth its beauty, there being no similarity, and therefore no rivalry between the forest trees and the blackberry bush.

It is not, however, an ordinary blackberry, but the very king of blackberries. It forms a round clump, some ten feet in height, and fifty in circumference, looking in the distance more like a haystack than a blackberry bush. It owes its form entirely to nature, and not to art; the long, flexible stems having sprung up, and, when they became too heavy to bear their own weight, having drooped over until they touched the ground. This has gone on year by year, each season adding to the size of the clump, and therefore to the length of the new stems. One stem of this year's growth is at least twenty feet in length, and at its base about as thick as a man's thumb. It has sprung from the middle of the clump, forced its way to the light, and then continued to grow so rapidly, when it could put forth its leaves, that it has overhung the whole of the clump, and its end lies trailing over the ground. Just in the same way, and for the same reason, the ferns in the New Forest run up to considerable heights, often exceeding eleven feet from the ground to their tips.

If the long trailing branches be drawn aside, the whole clump is seen to be constituted in a most curious manner. A mere shell of foliage and living shoots envelopes it, while the centre is composed of a thick tangled mass of dead, dry, and shrivelled branches, through which a few powerful shoots, such as that which has just been mentioned, have forced their way. Such a spot as this is just the place which rabbits love. All round the clump their resting places are at once discernible by a practised eye, the grass being pressed downwards in a sort of partial spiral, caused by the way in which the animals turn round and round before they lie down—a relic of wild life which has survived even in our domestic dogs. Many birds, by the way, make the foundation of their nests after a similar fashion, arranging a quantity of grass in the required place, sitting on it, and spinning round and round until they have worked a suitable hollow in it. Some thirty or so of these resting places are plainly seen, and while I was looking at the clump the other day, out popped a rabbit, without noticing me, as I was standing perfectly still. Suddenly the little animal detected a strange presence, and scuttled back again very much faster than it had come out.

As to position, this splendid blackberry bush is most favourably situated, being sheltered by trees on nearly every side, and only exposed towards the south. So much for the bush itself; now let us note its aspect in two periods of the autumn—namely, the beginning and towards the end.

It is a fine, warm, but not sultry day at the beginning of autumn, with a rather smart breeze, which is very pleasant in one respect, but rather disagreeable in another, inasmuch as it keeps at home so many insects which would otherwise be abroad.

As far as flowers and leaves go the bush is in its prime. Its whole surface is covered with flowers in various stages of progress. Some of them have nearly run their course; their petals are bent backwards, so

as to show their numerous yellow stamens, and the
nascent berry beneath them—rough, small, and giving
little promise of its future excellence. On the same
branches are very pale green, sharply-pointed buds,
that are yet unopened, while others have just broken
the green calyx, so as to show the pinky petals as they
lie folded within. Except the way in which the ample
gauzy wings of the earwig are folded under its tiny
wing-case, I know no natural packing so wonderful as
that of a flower while yet in its bud state. Even with
the small petals of the blackberry the arrangement of
the folds is worthy of attention ; but in large-petaled
flowers, which are often enclosed in comparatively
small buds, the complicated and yet simple disposition
of the folded petal is almost beyond the power of description—entirely beyond it without the aid of explanatory diagrams.

The leaves are all of the brightest green, with two
exceptions. On many of them are the tortuous tracks
of the tiny leaf-miner caterpillar, creatures so small
that they pass the whole of their lives between the
upper and under surfaces of the leaves, feeding on the
soft substance, called parenchyma, that lies between
them. Often they light on the toothed edge of the
leaf, and whenever they do so they seem unable to quit
the edge, though we think they would find more nutriment by following the example of many of their fellows,
and eating their way boldly into the middle of the leaf.
But like the mariners of old, who were always obliged

to keep near the shore, they were afraid to trust themselves in the open leaf, and followed every little projection with conscientious regularity.

Anyone who wants to know how gorgeous a British insect can be, let him tie a piece of gauze over some of these mined leaves, and watch them until a tiny moth appears in the gauze, the perfect state of the little caterpillar that made the zigzag track. It will be so small that it hardly looks like a moth at all, its outspread wings together not being as large as the letter O. Take the little creature to the microscope, concentrate a beam of light on it, and then, if you can, find words to describe its glories. It is useless to do so by comparing it with diamonds, sapphires, rubies, emeralds, gold, or so forth, because the moth infinitely surpasses the gems, and if we say that some imperial suite of jewels approaches the splendour of the leaf-mining moth, we should almost appear to be disparaging the insect. There are plenty of these wondrous little moths, every species having its own special beauties.

Some of the leaves are seen to be rolled up in little cylinders, while others are merely doubled, their edges being brought together and fastened. The difference of behaviour exhibited by the inhabitants of the rolled and folded leaves is very remarkable. If we pick one of the rolled leaves, out tumbles a little caterpillar, and drops to the ground, letting itself down by means of a silken thread, upon which it means to return to its dwelling when the danger is over. Pick one of the folded leaves, and the inhabitant gives no sign of alarm. Open it carefully, and there is the caterpillar inside, bright green, stretched out quite straight, and pressed so tightly against the central leaf-rib that at first sight it is not easy to discriminate between the insect and the leaf.

I noticed, on the same bush, two other curious instances of protective resemblance. One was that of a large tipula, one of those insects we collectively know as daddy long-legs. Whenever these flies settled on the bush, they invariably chose the flower-stalk as their resting-place, so that their legs, wings, and bodies became mixed up with the diverging lines of the flower-stalks; and if the eye were once taken from them they could hardly be detected.

The second instance was a much more singular one. Great numbers of the common garden-spiders had hung their wheel-shaped webs among the branches of the bush. One was a very beautiful spider, with a bright green body, so conspicuous, indeed, that I wondered how it could ever escape observation. So I gave the net a jerk, in order to alarm the spider. As soon as she felt the jerk she left the web, ran along a plant to the spot where two leaves sprang from the stem, plunged into the angle formed by their junction, and tucked away her legs under her. In this attitude her green body looked so exactly like a small leaf-bud that it was hardly possible to persuade myself that the creature was not a veritable bud. And the similitude was increased by a little red spot at the end of the tail, which exactly represented the red top of an unopened bud.

On the particular day which I have mentioned there was too much wind to please the insects, very few of whom, therefore, made their appearance on the blackberry bush. The hive-bees were tolerably busy among the flowers; as, indeed, they always are whenever they have a chance. Humble bees also came with their heavy hum, and accordingly, when a hive-bee and a humble-bee wanted to take to the same flower, there was a difference of opinion, which, however, never came even to the semblance of a fight. Several species of solitary bees also came, more apparently for the pollen than for honey.

Many ichneumon flies were fussing about among the foliage, but I have not yet been able to discover the object of their search. I suspect, however, that some of them at least may have come after the leaf-rolling and leaf-folding caterpillars. We know that there are some ichneumon flies which can even get at larvæ that are buried in the trunks of trees, boring through the solid wood with the long hair-like ovipositor, and contriving, by some wonderful instinct, to hit upon the very spot in which the larva lies hidden, and to pass an egg into its body along the ovipositor. So I thought that these ichneumons, especially those which had tolerably long ovipositors, might be on a similar errand, endeavouring to pierce the caterpillars within their leafy houses. I could not however detect one ichneumon so engaged, though I watched very carefully.

Several species of hoverer-fly were about the bush, settling occasionally on the flowers. These beautiful and strong-winged insects care little for wind, and may be seen enjoying their brilliant life on the wing when even the bees scarcely like to venture out of their hives. Some of them bear a curious resemblance to wasps, and others to bees, so closely imitating those insects, not only in shape and colour, but in action, that no one who is not a practical entomologist will venture to take one in his hand.

Of beetles there were but very few. Some of the tiny pear-shaped weevils were crawling slowly over the leaves, clinging to them very firmly, considering that they were sharply agitated by the wind. Some of the long-bodied, soft-skinned beetles, known as soldiers and sailors, according to their red or blue colour, were paying their respects to the blossoms. They are usually to be found on umbelliferous flowers, but there are few plants on which they may not be seen. They are rather quick on foot, and have a fashion of quivering their long antennæ in a manner exactly resembling that of the ichneumon flies.

There is a peculiar trait of character among these beetles. Being so soft-bodied, they seem little fit for combat, and yet there are no insects more apt to fight, or more ferocious in their battles. They are as fierce and quarrelsome as game-cocks, and, like them, will fight to the very death. But the game-cock when he has vanquished his opponent merely proclaims his victory aloud and leaves his antagonist, whether he be dead or disabled. The soldier and sailor beetles, however, are not satisfied with the death of their adversaries, but must need consummate their victory by eating their conquered foes. *Vœ victis* is their motto, and thoroughly is it carried out. If the doctrine of development be true, the Fijians must have been soldier beetles at some early stage of their progress towards humanity, and retained their custom of devouring their slain foes, though they have sadly deteriorated in point of courage.

Towards the end of Autumn.—The Blackberry bush now assumes a new aspect.

It is different in itself, its surroundings are different, and its visitants are different. The clump of distant trees on its western side gives many a sign that the foliage has passed its prime and has taken the first step towards decay. The leaves have darkened, and many of them show the brown tint of fading life round their edges, while the whole tree is studded with the green balls of the prickly fruit, some of which have burst and shed their richly coloured contents, while others hang unbroken on the boughs. The birch trees show bright yellow spots here and there among their foliage, the elms are dropping their leaves and are visibly more bare than they were a fortnight ago, while the background of bracken has lost its beautiful green, and is little more than a mass of yellow and brown, flecked sparely with green where a few of the younger plants still preserve their colour. Only the oaks remain apparently unchanged, but then they are always late, both in getting their foliage and losing it. Indeed, so tenacious are the leaves of their hold, that they cling to their branches throughout the winter, and only fall when the new leaves of the ensuing spring force them from their place.

But the greatest change of all has fallen on our blackberry bush. All its bright freshness has gone, and there is little to remind us of its vanished beauties. The leaves are dull, harsh, and brittle, and most of them are dimmed with many a spot and patch of black, yellow, or brown, while in many cases all three colours are to be found on the same leaf, a few straggling and uncertain dashes of the original green being left between them.

In fact, the leaves have now almost fulfilled their mission both to the plant and its visitants. The winter's repose is at hand, the plant will no more extend its growth; the rootlets which supply it with nourishment have served their office, and the leaves, which are its lungs, need no longer supply it with air. There is a wonderful and very close analogy between a hibernating animal and a tree in winter time. Indeed, the latter is truly a living creature, though on a lower plane than the lowest of the animals, and, as a partaker of life, it accepts the two conditions of life—nutriment and respiration. Let the creature, whether plant or animal, be able to exist for a time without the former of these conditions—or, rather, to exist for a time on a store of nutriment already laid up—and the latter condition may be almost in abeyance.

Take, for example, any of our hibernating animals, from the mammal to the insect, and see how slight and almost imperceptible is the respiration during the time that nourishment ceases. We need not take into consideration those insects which, in a perfect state, consume no nourishment whatever, and yet act and respire vigorously. Every one of them lives but a very short life. They are burning away the stores of fuel already laid up, and a few days at most are the utmost limit of their existence. They have just sufficient vital power to seek their mates and deposit their eggs, and straightway die. But the plant has a comparatively long life before it, and so has the hibernating animal; and therefore during the winter time there is vitality enough to enable the creature to revive itself when the season of spring comes round in its annual course. Specially is this the case with the tree. Battered, withered, pierced, torn, and half-eaten, the leaves of one year could never act as efficient respiratory organs for the increased needs of the tree in the ensuing season. So

MY BLACKBERRY BUSH.

the old leaves fall in the autumn of the fading year, only to be replaced by fresh and vigorous foliage in the coming spring, their multitudinous air passages un-choked and undamaged, and their myriad mouths wide open to receive the breath of life.

Look, for example, at those leaves of our blackberry bush, and see how utterly unfit they are for their proper work. There is scarcely one of them that is not in some way injured by causes external to it itself. The slime track over some gnawed and scalloped leaves tells that snail or slug, or both, has passed that way. On others the now transparent tracks of the leaf-miner caterpillar show that the respiratory value of that particular leaf is a thing of the past. Next comes one of the rolled or doubled leaves, abandoned long ago by its original maker, but never tenantless. Open one of these deserted habitations, and out scuttles a spider, or perhaps three or four earwigs, and now and then a woodlouse. If you wish to take a lesson in the art of packing, catch one of the earwigs, kill it by dashing it into boiling water, take it out at once, and unfold one of its wings. When every crease is laid open try to re-fold the wing, and put it back in its place. With two hands, a microscope, and unlimited instruments, you will scarcely achieve in two hours the task which the insect completes in two seconds, with nothing but its tail pincers by way of tools.

So much for the leaves at the end of autumn. As for the flowers, they are nearly all gone. A few—a

very few—still remain, to which assiduous attention is
being paid by sundry bees and hoverer flies, and all
these are on the sheltered side of the bush. Over the
whole of the clump the berries have taken the place of
the flowers. There is a curious lack of berries through
the lower parts of the bush, scarcely a respectable berry
being visible within four feet of the ground. This
phenomenon is soon explained, for I meet a couple of
fair golden-haired children, taking turns at driving a
wheelbarrow half full of blackberries, every one of which
has been picked from this special bush, and to every
one of which they are heartily welcome.

All the northern side of the bush is flecked with
large tufts of thistledown. The thistles themselves
are a long way off, growing on some neglected ground,
and the sharp north wind has stripped them of their
down, and whirled it along until it has been intercepted
by the prickly branches of the blackberry bush. That
the whole land is not overrun with thistles we are
indebted to the finches, the principal of which in use
and beauty is the goldfinch, flocks of which may be
seen flitting along the hedges or over the ground,
picking up the thistledown as it is whirled lightly
along by the wind.

There is still too much wind for insects. In this
part of the country at least, which is perched on the
top of a hill, the present season has not been favourable
for insects. When the weather has been fine and the
sun hot, there has been a smart breeze blowing; and

when the wind has fallen the sky has been cloudy, often raining, and the barometer has fallen. However, the very fact of the windy weather has brought out some insect faculties in a way that could not have been observed during a calm.

For example, the splendid peacock and red admiral butterflies whirl by us, making a circle or two round the bush as if tempted by its flowers or fruit—most likely the latter, and are then carried off by the breeze as if any number of blackberries were not worth the trouble of fighting the wind.

A deep ominous hum and a yellow streak in the air. Another hum and another yellow streak. These are hornets, the last survivors of their community. They have a nest somewhere in the grounds and are going straight to it. In a week or two more not a hornet will be found in the nest, though some females will survive in the winter, hung up somewhere, in bat fashion, by the claws of their hind legs, eating nothing, and scarcely breathing at all until the succeeding spring-tide releases them to a brief period of activity. Never a male hornet lives through the winter. He is not wanted and therefore does not exist in a world which tolerates no idlers.

Still the ichneumon flies are prowling about, and chief among them is that large, pale yellow species, with its long antennæ and sickle-shaped body, which is called, scientifically, *Ophion luteum*. Looking to the usual habits of ichneumons, it is rather a night flier than a denizen of day, and has a habit of getting into rooms after the lamps are lighted, flying into the flame, scorching itself, and falling on the table, where it spins about until some one mercifully puts it out of pain. It is rather a formidable insect to handle, behaving exactly as if it had a sting, and if touched, it brings the end of its body against the fingers just as a wasp or bee would do.

Presently there is a quick ruffling sound, and some large insect swoops past. It is a dragon-fly, which is out on a marauding excursion, taking the blackberry bush as its base of operations. It is curious to see how the movements of a dragon-fly in search of prey resemble those of a flycatcher. The bird takes a stand on some elevated spot, and then makes short flights into the air, catching an insect, and returning to its post. So does the dragon-fly. It picks out some suitable resting-place, and makes that the spot whence it issues in search of prey. It settles, waits for a moment, and then, as if a hidden spring were touched, down go all its wings with a jerk, and the creature remains motionless except its head, which is twisted from side to side in a way that is almost comical, while its great round eyes glow in the sunshine like a pair of opals. Suddenly, and without any previous preparation, off it darts into the air, makes a bold swoop or two, and then returns to the spot whence it started.

I noticed two peculiarities in its habits. In the first place, like the daddy long-legs, which has already been mentioned, the dragon-fly always contrived to make itself look exactly like the object on which it was settled. Its favourite post was a dried twig, from which projected a few withered leaf-stalks and parts of leaves. On this twig the insect took its station, sitting longitudinally, so that its slender body coincided with the outline of the twig, and its motionless wings looked wonderfully like the withered leaves. Not even its legs betrayed it, as they were not spread out, the tip of each foot taking a separate hold, as is generally the case with insects, but were all gathered closely together, like that of a goat when the animal is standing upon a narrow ledge.

In the next place it invariably darted into the air whenever there came a sharper gust of wind than usual. At first I thought that the insect had simply been blown from its hold, but then reflected that a dragon-fly is a powerful and tight-clinging insect, and not likely to allow itself to be blown from its foothold. Moreover, its station was on the sheltered side of the bush, where the effect of the gusty breeze was very trifling. After watching it for a time I found out the motive of its actions—a motive which points rather to reason than to instinct on the part of the dragon-fly.

As a rule, the insects which were on the wing were at such times unable to resist the sudden gusts of wind, and were whirled away without much power of directing their course. The firm, large, and powerful wings of the dragon-fly, however, were nearly independent of the wind, so that the insect of feeble flight was completely in its power—I was going to say, at its mercy, only that it has none. In fact, the dragon-fly has at such times an advantage over most other flying insects, much like that which is enjoyed by a steamer over a sailing vessel when both are making their way against contrary winds.

In this way insect after insect was captured, various species of frog-hoppers, by the way, appearing to form the staple of the dragon-fly's food. Their two white, milky, slight wings had no chance against the four swift wings of the dragon-fly, which swooped at them as they were blown along helplessly by the wind, caught them with unerring certainty, settled on its resting-place, gobbled them up with a couple of bites, just as a mastiff disposes of a mutton chop, and then looked hungrily round for more prey.

Let everyone who values the balance of Nature protect and encourage the dragon-flies as much as possible. They do no harm in any way, and they do an infinity of good by feeding upon insects, many of which are destructive either to the field, the garden, or the orchard.

THE REPOSE OF NATURE.

SIX months have passed since my readers took with me a 'Summer's Walk through a Country Lane.' The earth has since then accomplished nearly one half of its aërial course; and reader, author, and lane have traversed a space of some two hundred and seventy million miles, passed through the seasons of genial Summer, fruitful Autumn, and have commenced the cold Winter time, the season of the earth's repose. Our beautiful trees, with their heavy masses of varied green, have changed gradually from bright emerald to dark olive, and passed through successive phases of redundant colouring that defy the artist's brush to imitate, until they have finally settled down into ruddy brown and sombre grey. The leaves have fluttered one by one to the earth, which lies below waiting to receive their withered forms into her bosom. She waits to transmute these effete particles into new forms of life and beauty, and to cause a future progeny of young and vigorous leafage to spring Phœnix-like from the funeral pyre of their ancestors, spontaneously raised under the shadow of their parental tree, fired by the hot beams of the summer sun, fanned by the breezes of spring, and quenched by the rains of autumn and the snows of winter.

Our hedges are bare and scanty, with the bright light shining through their denuded gaplets that so recently were veiled with rich verdure and blossoming flowers; our path is hard, sharp, and treacherous, and our feet likely to slip from the frozen pebbles and deposit us in the ditch, lately so full of flowers, but now containing a mixture of snow, water, dead thorn-branches at the bottom, and a few thistle-stems and nettle-leaves on the sides, that render such a locality a singularly unpleasant sojourn. Even our dear little pond is covered with ice, except where a few persevering ducks have swum so continually round a tiny circle that the water still bubbles through the icy covering, and where the cattle have still managed to break away the frozen surface in order to drink, thereby kneading the water into a kind of muddy paste, and covering the neighbouring ice with most unsightly brown splashes. Our little streamlet is dry, and the many creatures that disport themselves in its rippling waves have disappeared.

Gone are the insect tribes, whose busy hum gave such life to the scene; not even a beetle is to be seen taking a short stroll from one tree-root to another; hardly a bird has enough spirit to utter its lively chirrup, and the very robin himself, with his brown coat and red waistcoat, has gone off to the farmyards and houses, trusting to his insinuating ways, his sly boldness, and the irresistible compassion excited by his pitiful aspect as he sits outside the windows, with ruffled feathers, sunken head, and bright eye gleaming from the downy plumes. The cunning little fellow seems to feel that no one sitting in a warm room, at an abundantly spread table, can resist opening the window and giving a hearty welcome to the 'ittle baird with boothom wed,' as one of my child-friends is accustomed to call him. So the window is raised, and in comes the feathered mendicant, at first shy and fearful, keeping at a respectful distance, and picking up the crumbs that are thrown to him, with many a sidelong hop and great flirting of the wings.

Compassionate reader, if your premises should be invaded by a poor, cold, half-starved robin, do not feed him with bread-crumbs, but give him some little bits of fat meat cut in long and thin strips like small worms. Of course he will eat the crumbs provided he can get nothing better, but he requires the meat to supply his glowing frame with the capability of resisting the chilling frost. He will not forget your kindness, but day by day will make his appearance at your window, hop about your table, eat out of your hand, and repay you with one of his own bright songs, which to my ears have the most charming mixture of mirth and melody.

There is, however, one drawback in his character. He is dreadfully jealous, and will not permit another bird to avail itself of the hospitality to which he has been indebted for his life, and has been known to kill in succession a whole series of unfortunate redbreasts that happened to trespass on the ground which he considered as his peculiar property.

Perhaps our lane is knee-deep in snow, and path, ditch, hedge, tree, and field are alike clothed with one uniform mantle of shining white, glittering here and there as the cold sunbeams sparkle on the sharp snow-crystals that gleam like microscopic jewellery from every spray.

Where are all the busy, merry creatures that flitted among the branches, traversed the soil, or urged their course through the waters? Some, such as certain migratory birds, have flown to warmer regions, many have perished with the first frosts, having completed their earthly mission, while myriad others are still living in some recess, quiescent to all external appearance, but full of life and activity within, either sunk in that marvellous state of existence which seems really to be half-way between sleep and death, or undergoing a total change of being, in readiness for the ensuing spring.

We miss our little friend, the squirrel, from his accustomed haunts. No longer is he to be seen scudding about the grass in his own odd fashion, squatting upright with his feathery tail curled parasol-wise over his head, picking up a beech-nut with his fore-paws, nibbling at it critically, and then throwing it away and hopping after another. No longer can we amuse ourselves by rushing at him suddenly, and seeing him go leaping over the ground with his brush trailing behind him, and his body looking double its real length; watch him jump at the trunk of a tree, slip round the stem, scud up the branches, and then sit coolly on the topmost bough, and look down at us with benignant disdain.

Our little friend has gone to sleep for the winter, and if you know where to find his 'cage,' you may catch him asleep without much difficulty. Be it remembered that he has two homes, a summer and a winter house; the former being lodged in the fork of some lofty branch, often near the end of a slight bough, and very conspicuous from below, and the latter warmly established close to the trunk of a goodly tree, sheltered from chilling winds by the large limbs against which it is placed, and defended from rain and storm by the well-thatched roof and warm lining.

Snugly coiled in this warm recess the squirrel passes his winter, spending very many consecutive hours in that strange sleep which is called hibernation; awaking at intervals, when a gleam of warmer sunshine than usual rests upon his cage, running to his hidden treasury, taking a little refreshment, and then returning to his house to fall asleep again. He has an excellent memory, this little squirrel, and his faculties are not at all beclouded by the long hours of sleep; for as soon as he wakes he comes quietly out of his warm cottage, scrambles down the tree, runs to one of the spots where he has laid up a store of food, scratches away until he has disclosed his treasure of nuts, takes as much as he needs, and returns to his home. Even when the snow lies thickly on the ground he is at no loss, but, guided by some intuitive power, proceeds to the spot with unerring certainty, scrapes away the snow, and secures his meal.

The squirrel has a distant relation, a kind of third cousin once removed, well known under the title dormouse, and often seen in cages, but not very frequently in a wild state. This little creature is also one of the hibernators, and has its warm nest in a thick bush, much as the squirrel has its domicile in a tree, where it sleeps its time away throughout the winter. Like the squirrel, too, it has its store of food, not gathered into the earth, but tucked away into sundry nooks and crannies in the neighbourhood. The amount of food which the dormouse takes during the winter, and the frequency of its awakening, depend almost entirely on the severity or mildness of the season. In a very sharp winter the drowsy creature wakes but seldom, and very little of its store is consumed, and indeed, even should the season be mild, the inroads on the larder are but few. The provisions are not gathered so much for the winter as for the first few weeks of spring, when the animal has at last shaken off its long wintry sleep, and returns to its own lively habits, nature not yet having supplied it with a sufficiency of food whereon to live.

The hedgehog, too, is another of our hibernating animals, coiling itself up in a warm nest in some hollow tree, or under the gnarled and projecting roots, and occasionally seeking a domicile in a deserted rabbit-burrow or disused fox-hole. All these three creatures may be found sleeping in their homes, and are thus easily captured.

As all these animals awake at intervals during the winter, and partake of nourishment, they are said to be partial hibernators, the best British examples of perfect hibernation being exhibited by those singular winged quadrupeds which we call bats.

If in winter we explore the recesses of almost any hollow tree, any dark crevice in the rocks, or any old deserted building, there we shall find, hanging by their hind legs or gathered closely into thick clusters, some bats, sunk in the deepest lethargy, and giving but slight indications of life. All through the winter hang the bats, with scarcely a movement of head or limb, and, unlike the preceding animals, they never awake to seek nourishment, as there would be none for them.

There seems to be no creature which spends so much of its time in sleep as the bat. Not only does it lie dormant throughout the winter, but it passes daily into that strange state of drowsiness which is more than sleep, though not quite so deep as in winter.

It is a popular but very erroneous notion, that this torpor is caused by cold. Now, if this were the case, the hibernating animals would place themselves in some cold and exposed spot, where they would be influenced by the increasing chill of the weather. But it is found, after a long course of experiments, including a most valuable series by Dr. Marshall Hall, that the effect of cold upon a hibernating animal is twofold; it first awakes the creature from slumber, and then kills it.

During the time of its slumbers the extreme torpidity of the vital organs is most curious, while the external portions seem to acquire a proportionate irritability, a phenomenon which is partially seen even in ourselves during ordinary sleep. If, for example, a hedgehog while in the torpid state be touched, it partially uncoils, gives a peculiar deep grunt, and again curls itself up. The bat, if touched while in this strange sleep, will wriggle about like an injured worm, while the very same touch would have no perceptible effect upon it when awake. Indeed, the hibernating creature seems to pass, for a time, into a lower state of being, as far as its mere animal characteristics are concerned; and the bat, the highest of our British mammals, becomes scarcely higher in its organisation than a toad or a frog.

Instead of keeping up a high temperature, as is the case while it is awake, it actually becomes colder than many cold-blooded animals; the temperature of the body exactly following that of the surrounding atmosphere, the heat of the surface being about half a degree higher, and that of the vital organs about three degrees; so that when a thermometer hanging beside the animal marked a temperature of thirty-six degrees, one whose bulb was within the stomach only marked thirty-nine. A similar phenomenon was observed by Dr. Jenner with a hedgehog, when the internal temperature was a little over three degrees above that of the air.

The reader must bear in mind that experiments of this nature require the very greatest care, for the hibernating state is of so delicate a nature, and so easily disturbed, that a heedless footstep on the floor will awaken the creature, set it breathing, and increase the temperature some twenty degrees in a minute or two.

During the true hibernation the breath is almost entirely suspended. Bats while sleeping have been gently immersed in water kept carefully at the same temperature as their bodies, have been sunk below the surface for a space of sixteen minutes, and found to be none the worse for their bath. A hedgehog has been subjected to the same test for more than twenty minutes, and although it moved slightly under water, and expelled a little air from his lungs, it was not at all injured by the experiment.

Hibernating animals have also been placed in carbonic acid gas for a space of several hours, without suffering from its effects, while rats and sparrows placed in the same gas fell lifeless to the bottom of the vessel and did not recover.

In order to ascertain with accuracy the rate of respiration and pulsation during this curious state, a hibernating bat was placed in an ingenious instrument which measures the quantity of air consumed by the creature contained within it. After remaining twenty-four hours in this machine, scientifically termed a pneumatometer, the index gave no sign. The animal was then slightly disturbed, and in the space of nearly three hours consumed about one cubic inch of oxygen. But when more disturbed, and forced to move briskly, it consumed five cubic inches in one hour.

Although the respiration is thus checked, and the lungs cease their labours, the heart continues to pulsate, though slowly, making less than thirty beats in a minute. So we have a most curious phenomenon, *i.e.*, blood constantly circulating through the system without any respiration to renew its vitality, and without even the reservoirs of air which are possessed by the reptiles.

The reason of the long hibernation of the bats is evident. They feed wholly on insects, which likewise disappear during the winter months; and if there were no means of reducing the bodily functions to the lowest ebb compatible with the retention of life within the frame, the whole race of insect-eating bats would be swept off the earth in a single winter.

It is sufficiently remarkable that the animals which hibernate on account of the absence of food should belong to the two extremes of the vertebrate kingdom. The squirrel and dormouse might lay up a store so large as to afford an abundant supply throughout the whole winter; but the bat, feeding only on animal substances, could not do so, and would starve but for the merciful torpor in which it is sunk for so long a time. For precisely the same reason our British reptiles retire during the cold months of the year into some deep recess, and there remain torpid until the succeeding spring brings with it the needful warmth and food.

Take, for example, two of our best-known reptiles, the frog and the snake, both of which disappear during winter for much the same reason. The frog lives chiefly on insects, which all vanish in the winter months for want of their vegetable food; and the snake also retires to winter quarters because it lives mostly on frogs, which have hidden themselves until the spring. Both these creatures, in common with many other reptiles, burrow deeply in the earth or seek some snug recess as soon as the autumn draws to a close, and, safe in their homes, sink to sleep until the sunbeams recover their warmth, again enliven the earth with verdure, and the annual resurrection of the vegetable world has been accomplished.

Then the renewal of that process takes place. The plants put forth their tender shoots, the leaf-eating insects come from their winter quarters to eat the leaves, the frogs emerge from the ground to eat the insects, and the snakes glide out to eat the frogs. Such reptiles as the blind-worm, which feed not upon frogs, but live on insects, slugs, and such-like creatures, are earlier than the snakes, because they find their food ready for them. Truly is it said, ' The eyes of all wait upon Thee, O Lord, and Thou givest them their meat in due season. Thou openest Thine hand, and fillest all things living with plenteousness.'

Strange discoveries are sometimes made in the course of gardening operations, if people will only use their eyes. A few years ago, while some workmen were pecking up the gravel in a playground belonging to a school at Oxford, preparatory to making certain alterations, they came on a little colony of frogs, about seven or eight inches below the surface, all sitting packed closely together, and all with their noses pointing to the surface. How long they had been in that situation I could not discover; but by comparing one circumstance with another, I came to the conclusion that the frogs had settled themselves down for their winter's slumber about two years previous to their disinterment, been covered with gravel when the playground was laid down, and had remained there perforce ever since. They were so firmly imbedded in the earth that they could not stir a limb, and must have depended wholly for respiration and subsistence on the small modicum of atmosphere and the very few insects that might make their way through the minute crevices which exist in all soil. In general the winter's retreat of the frog is in the muddy soil at the bottom of some pool or ditch, where they congregate closely together in masses, and remain without need of food or respiration until the spring.

In 1857 I was walking in the grounds of a gentleman living near Oxford, who was making considerable alterations in his domains. Observing a cavity rather curiously hollowed in a bank that was evidently being broken down for removal, I asked a man who was working near at hand if he knew the cause of so odd a piece of work. He told me that on the previous day he was cutting down the bank, when he came upon several large stones, and on removing them, he found a whole mass of snakes, tightly coiled up together, and closely filling the cavity in which they lay.

The hollow was about three feet from the surface of the ground, and, as far as I could make out from the aspect of the spot and the *débris* left by the workman, must have been about four or five feet from the face of the bank. I could not ascertain whether any aperture was visible, or any channel of communication between the hole where the snakes were found and the open air. The man, of course, thought they were vipers, agreeably to the invariable tendency of the rustic mind, which dreads the newt and the lizard, which are totally harmless, more than the viper, which really possesses a terrible store of poisoned weapons, and attributes to the bright and innocuous dragon-flies a sting worse than that of the wasp and hornet.

It has been already mentioned that a very great proportion of the insect tribes which buzzed and hummed so merrily during our summer walk have died after providing for a numerous progeny. Such is indeed the case; but there are many insects which are in full life, though at present in a state of partial torpidity, as is needful while the frozen ground and withered foliage afford them no sustenance.

There are the bees, for example, all snugly asleep in their hive, having contrived to keep up a sufficient warmth for their winter's needs, and laid up a sufficient store of honey for the little nourishment which they require. If you could look into their hives you would see the bees closely clustered together, and every un-sealed cell containing a bee that has crept half-way into it, and there lies comfortably sleeping. That cold is injurious you can easily prove by gently tapping the hive, when a little commotion is heard within, a bee comes out to see what is the matter, and immediately falls dead from the frosty atmosphere. Only do not repeat this process, unless you desire to lose all the bees—for when these insects awake they *must* eat, and unless they are kept perfectly quiet they will rapidly consume their store, and then die miserably of cold and hunger.

A few wasps, too, and other insects, may be found in banks or similar localities, there awaiting the spring, which will set them at liberty to initiate now house-holds and multiply their species in a marvellously rapid manner. The ant tribe too are patiently resting in their subterranean beds, and will be amongst the first to arise in the Spring.

Few persons have any idea, as they walk in the country on a winter day, how the ground beneath their feet is teeming with life. Putting aside the earthworms, and such creatures as have their normal exist-ence below the soil, we will just look for ourselves, and try to discover a few of the hidden wonders of this most wonderful earth.

Let us come to the feet of these elm and oak trees that are planted on the bank of our lane, clear away the snow, and begin to dig. In this sharp frosty weather, we shall need the aid of a pickaxe or some such weapon to pierce the frozen soil, but after the first few strokes a trowel, or even a pocket-knife, will answer tolerably well.

The best way to dig for insects is to peck up a circular patch about eighteen inches in diameter, throw aside the frozen clods, and then to work carefully down-wards, so as to form a conical depression in the soil. We shall hardly have dug four or five inches in depth when we shall come to our hidden friends. A big cocktail beetle is suddenly dislodged, rolls black and bewildered to the bottom of the hole, picks himself up again, runs at the supposed foe with open jaws, and defiant tail curled scorpionwise over his sooty back—falters, stops, runs on again, but slowly, as if paralysed—stops again, staggers, falls over and rolls back dead. He has been killed by the frost, because he was roused suddenly from his torpor.

Two or three more beetles of different species come tumbling out, and all meet the same fate, though not so dauntlessly as the cocktail. Presently we toss out, together with the mould, a brown spindle-shaped object, blunt at one end, sharply pointed at the other, and boldly ringed for half its length. This is the pupa, or chrysalis, of some large moth, and while removing it we lay our hands on one of the great mysteries of this world—a mystery, which, if rightly explained, would give the clue to many a bright truth now hidden within labyrinthian doubts and hazy theories.

At the very outset we are met with a paradox. The frost killed the beetle that came from precisely the same locality, and, of course, we might argue that this creature would also die from sudden exposure to the cold atmosphere. Nothing of the kind. Provided we do not handle it roughly, we may take it home, put it in a box, and in due time be rewarded by seeing a grand wide-winged moth emerge from the dull case in which it had so long lain, having suffered no injury from its unexpected change of residence. The more we dig, the greater number of living insects and pupæ hall we find, the former soon dying from the sudden cold, and the latter suffering no apparent inconvenience.

Here we have a totally different branch of the subject. What manner of state is this in which the chrysalis apparently reposes? It is not sleep, neither is it hibernation, but something quite distinct from both, and yet having a certain analogy to both. It is not death, for the creature still lives, and yet it is a kind of death to the caterpillar, which lately traversed the branches and fed on the green leaves, before it descended to a grave beneath the tree on which it had lived. It is not repose, for the vital powers are acting with wondrous force, vehemence, and rapidity, transmuting the heavy green caterpillar into the airy-winged moth, or rather evolving the one from the other, through the intermediate form which now lies dull, helpless, and apparently dead in our hands. Mystery of mysteries, all is mystery—unexplained, though perhaps not inexplicable—fraught, let us be sure, with wondrous meanings, and waiting until He who poured them from His all-creative being shall interpret their hidden prophecies!

I have called this article the Repose of Nature, for want of a better word; but, in truth, there is no absolute repose in nature. All nature rebels against it, and the powers of nature never cease from their labours. 'My Father works,' said the Lord, 'and I work;' and this is the law of the universe, operating on all created things alike. I fancy that there is nothing so abhorrent to the Great Worker as idleness—the pioneer of all picking and stealing, evil speaking, lying, and slandering. There is something within us which forces us to acknowledge the majesty of work; and the idlest man living can but feel an involuntary respect for the poorest industrious labourer who has died at his work, and a pang of remorse at the contrast to his own useless life.

All workers know that the truest rest is a change

of occupation, and that to be condemned to utter idleness would be the most terrible punishment that could be inflicted upon a human being. Why, even the poor fashionable idler really works, in his way, as hard as any of us, because getting amusement is much more laborious than getting a living, becomes more difficult every day, and leaves nothing but disappointment behind it. Idle people are fond of talking as if they had exhausted the world, and found it to be hollow and empty—like that poor silly man, of whom we read the other day in the papers, who shot himself because he had been all over this world and thought it was time for him to try another. Why, there is a sliver of a cedar-pencil lying on my paper, and I will answer for it that any ' used up ' personage who thinks that he has exhausted the world and will just try to find out all about that little slip of juniper wood will find life too short for the task.

Look, for example, at the amount of work which is achieved within this chrysalis lying before us, and just think of the millions upon millions of similar beings at this moment undergoing as complete a transformation, from a terrestrial to an aërial state of existence; their form, constitution, organisation, wants, and habits so totally changed that the one is wholly unrecognisable from the other. Even to go back for a moment to our old friend, the frog, what a wonderful law it is which takes possession of the no-limbed, long-tailed, gill-breathing tadpole, and changes it into a four-legged, leaping, air-breathing animal, without even a vestige of tail !

I know few pursuits more absorbingly interesting than tracing the gradual change of a larva or caterpillar while passing through its various states until it attains its perfected form, from which it never after varies. It is an easy task enough, and may be accomplished by anyone who has, or who chooses to acquire, a steady hand and a tolerable eye. Take any common caterpillars of rather large size—silkworms will answer the purpose well, and can easily be obtained—put two or three into proof spirits, and let the others change into their pupal form. Note the day that they change, and put a few into spirits within an hour after their casting off their caterpillar skin. Keep the rest, and every two days put a couple into spirits until the moths appear from the survivors, and then treat them after the same fashion. You will then have a really valuable series of objects, which by careful dissection under water or spirits will unveil some great mysteries. It is needful that the very early pupæ should be kept in the spirits for some weeks before dissection, as their interior is so soft as to be little but a milky fluid, and requires hardening with the spirits before it can safely be touched.

It is most wonderful to see the gradual development of the process by which a moth or butterfly is evolved from the caterpillar; the leaf-eating creature with its powerful jaws and huge stomach becoming a honey-sucker, with the most delicate digestive organs imaginable; the creeping thing changed into a winged being; the nearly blind grub into a creature with eyes of wonderful complexity; and the whole form of body, muscular system, nerves, and internal structure, being totally changed to suit the altered condition in which the remainder of its life will be spent.

Take, for example, the chrysalis which we have just dug out of the ground, and suppose the brown outer skin to be transparent while the process of evolvement is going on. During its caterpillar state nearly the whole of its body is filled with a huge stomach, extending throughout the greater part of its length, and tightly filled with food, as is likely in a creature that is always eating. The skin, which to the mass of spectators seems to contain nothing but a soft pulp, is lined with an array of flat and white muscles, and the whole space between these muscles and the stomach is filled up with fat, formed into rather hard lumps of variable dimensions, and penetrated with the breathing tubes, and some very slight nerves. Along the abdomen, and just below the skin, runs a chain of little knots of nerve-like substance, connected together with double cords of similar material; and along the back lies a chain of valves, which is analogous to the heart of the higher animals.

Throughout the transformation, the digestive, nervous, and circulating systems retain their relative positions, but are greatly altered in relative size and importance. The digestive organs are reduced to a tithe of their former volume, the masses of loose fat gradually shrink, while new members begin to make their appearance, and increase imperceptibly from day to day, gaining form and substance by the slow but Divine and irresistible power which is equally exerted in creating an universe or moulding a moth's plumage.

Mine ancient and constant enemy, lack of space, here warns me that we shall not be able to examine the whole structure of the future moth, and we will therefore restrict ourselves to the most obvious points of difference between the caterpillar and the perfect insect, namely, the wings. Under the skin of the back (and these can be seen even in the caterpillar) are two little projections, white, soft, and in shape not unlike the two halves of a pea, but rather flatter. On raising them with a needle it is found that each separates into two portions; and, on further examination, we find they are the latent wings in their unformed condition.

It seems incredible that within this little space should be packed the beautiful wings which, when spread, will contain several square inches of firm and strong membrane, penetrated by air-cells, strengthened by nervures, and clothed with myriads upon myriads of delicately carved scales. Yet it is the fact ; and, when the creature emerges from its case, we shall see how the wings attain their full size.

When the moth leaves the chrysalis state, it crawls up some perpendicular object, generally the native tree on which it has lived, and at whose foot it has burrowed. It then takes several long and deep inspirations, which have a perceptible effect in shaking out, as it were, the hair-like plumage of the body, and causing it to assume a brighter tint. It next slightly opens the wings, which are still thick and solid, and totally useless for flight. and communicates to them a rapid tremulous motion, every now and then pausing to take a few deep breaths. As it proceeds with this task, fold after fold is gently shaken and smoothed out, each breath driving the air through the tubes, which permeate every part of the wings, and so strengthening these members by regular degrees, until at last they stand out in all their beauty —firm, strong, and. pointed, and covered with a blazonry more gorgeous than ever herald (except the herald moth) endued.

Touch with a camel's-hair brush any part of the wing so as to remove a few scales, dab the brush on a slip of glass, put it under the microscope, and then see how each particle of the almost imperceptible and impalpable coloured dust which clothes the wings becomes manifest as an elegantly formed scale, sculptured with designs of singular beauty and regularity, formed of at least two, if not three, separate membranes, and waved, toothed, or fringed at the extremity, according to its position on the wing. Just consider how many hundreds of thousands of these scales are needed to cover a surface so great, and the inconceivable care which is required, not only in making them, but in setting them in rows more regular than the slates on a house-top, each over-lapping the other, and arranged so as to defend the delicate membrane of the wing from moisture. You cannot wet a moth's wing with water, for it runs off in drops as if the wings were covered with oil.

When were these scales made, and how were they fashioned? No naturalist can give an answer, save that they exist by the will of the Divine author. Truly it is worth while to reflect upon the constant and elaborate providential care which is required to form the wing of a moth in so short a time, and to think what laborious tasks are being elaborated in the earth beneath our feet, while we superficially think that nature is reposing. Not even the trees are reposing, although their branches wave, black and deathlike, against the sky. They are silently but laboriously concentrating their forces, settling the spots whence new leaves and new branches are to spring, driving fresh rootlets through the soil, in order to gather from its various elements those particles which will be needed to carry on the work of increase, and preparing themselves with the instinctive foresight of the vegetable kingdom for the labours of the ensuing year.

Even in so-called inorganic particles there is no absolute repose; for the chemist can detect in each grain of sand below our feet, in each tiny mite that dances and sparkles in the sunbeams, an array of mighty forces acting together, and uniting for the

time to preserve the object in the form which it at
present holds, but liable to be set free in a thousand
different ways, and then diverging upon their various
missions to do the will of the All-Worker.

He never slumbers nor sleeps; and hence it follows
that as all existences proceed from Him, as all created
things begin and end in Him, all things must necessarily be imbued with the spirit of eternal and ceaseless labour; and, though they may for a while rest
from their labours—and "their works do follow them"—
can never suffer stagnation, and much less be annihilated. Each material particle which assists in the
constitution or the functions of our mortal bodies
brooks not stagnation for an instant, but with a
curious and evident analogy passes from death to life,
and becomes etherealised into its most rarefied and
gaseous forms, another being, and yet the' same !

TURKEY AND OYSTERS.

MANY affinities lie dormant in Nature.

How incredulous would have been the ancient Briton in his light costume of woad, and the aboriginal American in his war paint, had a Druid or a medicine-man foretold that their far distant countries would be linked together in gastronomic bonds, and that the turkey and the oyster would be ever associated in the minds of a future posterity! How their real affinity was discovered is a problem as yet unsolved, and too closely interwoven with the progress of the human race to be examined in any work of less dimensions than a folio. But the fact is patent, and henceforth the turkey and the oyster are wedded together as indissolubly as the bacon and beans of the rustic, the whitebait and lemon-juice of the cabinet minister, and the chops and tomato sauce of Mr. Pickwick.

I may be justified in supposing that in every household where this essay will be read—that is to say, in every respectable household throughout the kingdom—a hamper containing a turkey and a barrel of oysters has either been received or sent as a present elsewhere at Christmas. Sometimes both events occur simultaneously, and the same P. D. C. cart which takes away a hamper containing a turkey and a barrel of oysters, deposits another hamper containing another turkey and another barrel of oysters. An enquiring mind cannot but be struck with the enormous multitudes of birds and molluscs that must be bred in order to supply even the vast annual demand for Christmas, taking no account of those that are consumed during the other seasons of the year.

To begin at the beginning. For the first knowledge of the turkey we are indebted to Columbus, inasmuch as the bird is indigenous to America, and is by no means a native of Turkey, as is the general but mistaken idea. The popular name was given to the bird in allusion to its proud and haughty strut, its unconscionably large harem, and its irascible temper. For at the time when the bird was first brought into notice the Turks were a dominant nation, with rather more than the usual intolerant arrogance which is likely to characterise a people at once powerful, bigoted, ignorant, and exclusive. Even at the present day, when the once all-powerful nation has sunk into the position of a mere province, whose very existence is only maintained by the common consent of surrounding countries, the regular orthodox Turk is as supremely contemptuous towards an infidel as in the days of his ascendancy, though he dares not express his feelings except by low and muttered curses. As it is, he will seize every favourable opportunity of applying the epithets of dog and pig to casual unbelievers, and express the very lowest opinion of their female relatives, so that in the time of his power his arrogance must have been just unbearable.

In our farm-yards, where the turkey knows his place, and is subdued unto domesticity, he behaves in a very different manner from the free wild bird in his native woods, who lowers his crested head for none, who rules with undisputed sway over his female train, and has won his way to eminence by successive victories. He is a grand bird and a proud one, as he stalks majestically through the woods followed by his obedient troop, like a patriarch of old with his wives and children; ruffles his feathers and spreads his tail in sheer exuberance of pride, and ever and anon gives vent to that extraordinary sound which we call gobbling, and which Arabs have mistaken for a dialect of their own guttural language.

At the present day the turkey is a potent ally to those far-seeing enquirers who are giving their best endeavours to enrich this country by acclimatising the useful denizens of other lands. There are many most valuable creatures—beasts, birds, and fishes—which are gradually being 'improved' off the face of the earth, and which, unless we grant them a resting-place, will in a few years be as extinct as the mammoth, the dodo, or the iguanodon.

The progress of civilisation is rapidly producing its

effect even upon the turkey. In former days the wild turkey wandered in vast multitudes throughout the northern parts of the United States, suffering but little harm from the American Indian, and fearful only of the natural enemies to which every wild being is subject. Now, however, all is changed. First the pioneers pushed their way into the interior; then the squatters raised their log huts and made their 'tomahawk improvements;' next came the settlers, each house forming the centre of an ever-enlarging circle, within which no beast could venture without imminent risk of death. Villages sprang from settlements, cities grew out of villages, and man took undisputed possession of the territory that was no longer a home for game.

A recent writer on American sports states that the wild turkey is slowly but surely perishing. Few or none are now to be seen north or east of Pennsylvania, and only a very few in some of the remotest parts of that State. In the wildest parts of Virginia a few families yet linger, but they increase in number towards Ohio, Indiana, Illinois, and Kentucky. Those who wish to see this noble bird in perfection must go to Arkansas, Louisiana, Alabama, Mississippi, and Tennessee—and even then, they will have to be well skilled in the hunter's craft before they will come within sight of the wary bird. Fortunately the turkey has been acclimatised in many countries, so that there is little real danger of its entire extinction. But though captivity does not destroy the race, it sadly dims their colours, and in the third generation at the farthest the brilliant purple lustre and glossy metallic bronze of the back and wings, and the rich green, bright chestnut, and velvety black of many feathers have sobered into brown ochre, and dull sooty black.

Of course there are plenty of opponents who deny the whole scheme as an .impossibility. Some assert that no animal will prosper except in its own land, and that all imported specimens will die out unless recruited by fresh arrivals. Now the turkey happens to be one of the very beings whose existence is thus denied. We want no importation of wild turkeys from America in order to add vigour to our domesticated specimens, and the possibility of acclimatisation is thus triumphantly proved. On the contrary, their habits are already too wild; they are terrible rovers, require to be watched as carefully as a sentinel watches his prisoners, and employ every device in order to escape from constraint. A hen turkey, for example, always tries to steal away just before laying, and establishes her nest in some spot so well concealed that it frequently escapes all the sharp eyes that have been searching for it. It is by no means a rare occurrence for a hen turkey to be suddenly missed from the farm-yard, and after some weeks have elapsed to return with perfect composure, leading a whole train of young chicks behind her.

The turkey has a great objection to confinement, and is a very gipsy in its love for open air. If possible, it always roosts in a tree, and prefers to sit on the branches during the coldest night, rather than rest warmly in a comfortable shed. It is a charming bird on the table, whether roast or boiled, but it gives great trouble in the field. It loves to roam about and pick up the insects, seeds, and other food that it may light upon in the course of its rambles. It has a special liking for traversing hedge-rows, and will spend hour after hour in this pursuit, never seeming to weary, and pecking away as smartly at the end as at the beginning of its run. The only method of securing the return of the turkey is to make a practice of feeding it well in the evening, choosing some diet of which it is especially fond. It is then sure to come home and partake of the food, and can be quietly shut up while discussing the viands.

Though a native of Northern America, and subject therefore to extreme cold, it does not seem to bear our comparatively mild winter when young, and is especially sensitive to water, being apt to die if wetted. After they have passed through their chickenhood the young birds are much more hardy, and require less care. In mere point of hardihood they are equal to any of our indigenous birds, provided that they have fairly attained their maturity. They can endure a severe frosty night, spent in the open air, without apparent inconvenience, even though their feet should be frozen to the branches on which they have perched. But they are always perilous creatures to manage, and will not repay their owner for his trouble, unless he takes pains to acquaint himself with their habits.

One of the most dangerous periods is that wherein the distinctive marks of the two sexes begin to appear. The chicks require plenty of nourishing food during the day, and must be carefully housed at night. As soon, however, as the wattle on the forehead and the wrinkled skin of the neck show themselves the danger is considered as past. Then they *will* roost on the topmost branches of trees, if they can manage to escape from the watchful eyes of their keeper, a habit inherited from their ancestors, who always perched in trees at night in order to escape from the lynx and other rapacious animals that prey upon these delicate-flavoured birds. A whole flock will sometimes fly into a tree, and when once among the branches will not come down again. Altogether, they are restless, wandering birds, and unless they are watched with the greatest care they are sure to fail. Care, however, is the one great essential in rearing these magnificent poultry, and even in the most unfavourable parts of England flocks of turkeys have been bred which will bear comparison with the best specimens produced in Norfolk, the chief county for these birds.

Now let us change our theme and pass to the oyster, the natural companion of the turkey.

Even in the remote ages of the world, when Rome was in the ascendant, the mistress of the globe, when our isles were deemed to be the extremest boundary of the habitable world, and the fit home for pestilence and disease, that had been driven by the power of the gods from the City of the Seven Hills, Britain was yet of some importance to the civilised world. She produced the oyster. 'Natives' are no modern delicacy. Lucullus always had them on his table. They were set before emperors, and devoured by certain imperial gluttons in vast quantities that even surpassed the feats of their modern imitator, whose name was, at the beginning of the present century, as familiar in all men's mouths as the oyster was in his own. If, too, we may judge by the confessions of Christopher North, oysters were consumed in the 'Noctes Ambrosianæ' with as much fervour as in the ancient times, inasmuch as each member of the famous trio seemed to consider himself hardly used if he only had two hundred oysters by way of getting an appetite for the supper that was to follow.

There is certainly something about a barrel of oysters that wears a most fascinating aspect. For my own part, I can hardly conceive a more luxurious entertainment than to have a whole long winter's evening to myself, with the unwonted feeling of nothing to do, slippers, a bright fire, which will not smoke on any provocation, unlimited Cobb's ale, a fresh barrel of natives, and a vision of egg-flip to follow.

As to such heresies as pepper and vinegar, let them be banished from the table whilst oysters are upon it. These charming molluscs should always be taken unmitigated, without losing the delicacy of their flavour by a mixture with any condiment whatever, except their native juice. Alas! there are but few who know how to appreciate and make use of these natural advantages. Scarcely one man in a thousand knows how to open an oyster, and less how to eat it. The ordinary system which is employed at the oyster shops is radically false, for all the juice is lost, and the oyster is left to become dry and insipid upon the flat shell, which effectually answers as a drain to convey off the liquid, which is to the oyster what the 'milk' is to the cocoa-nut.

Those who wish to eat oysters as they should be eaten should act as follows :—

Hold the mollusc firmly in a cloth, insert the point of the knife neatly just before the edge of the upper shell, give a quick, decided pressure until the point is felt to glide along the polished inner surface of the under shell. Force it sharply to the hinge, give a smart wrench rather towards the right hand, and off comes the shell. Then pass the knife quickly under the oyster, separate it from its attachment, let it fall into the lower shell, floating in the juice, lift it quickly to the lips, and eat it before the delicate aroma has been dissipated into the atmosphere. There is as much difference between an oyster thus opened and eaten, as between champagne frothing and leaping out of the silver-necked bottle, and the same wine after it has been allowed to stand for six hours with the cork removed.

There is another method of eating oysters, wherein no knife is required and not the least skill in opening is needed, the only instrument being a pair of tongs, and the only requisite being a bright fire. You pick out a glowing spot in the fire, where there are no flames and no black pieces of coal to dart jets of smoke exactly in the place where they are not wanted, as always takes place during the operation of making toast. You then insert a row of oysters into the glowing coals, taking care to keep their mouths outwards, and within an easy grasp of the tongs, and their convexity downwards. Presently a spitting and hissing sound is heard, which gradually increases until the shells begin to open, and the juice is seen boiling merrily within, the mollusc itself becoming whiter and more opaque as the operation continues. There is no rule for ascertaining the precise point at which the cooking is completed, for everyone has his own taste, and must learn by personal experience. A little practice soon makes perfect, and the expert operator will be able to keep up a continual supply as fast as he can manage to eat them. When they are thoroughly cooked they should be taken from the fire, a second batch inserted, and the still hissing and sputtering molluscs be eaten ' screeching' hot.

A true ostreophilist will never eat oysters in any but one of these two methods, and holds that in oyster sauce, oyster patties, scalloped oysters, and the many other dishes in which these bivalves are employed, the oyster is wasted, and the accessories might have been turned to better account. No one who has not eaten oysters dressed in this primitive mode has the least idea of the piquant flavour of which they are capable. Stewed in their own juice, the action of fire only brings out the full flavour, and as the juice is consumed as well as the oyster there is no waste, and no dissipation of the indescribable but potent aroma.

The immediate contact of fire, the great purifying and vivifying influence of the material world, has a wondrous effect upon the objects submitted to its influence. There should be as few intervening substances as possible between the fire and the food. Are not chops and steaks broiled over glowing charcoal infinitely superior to the very same viands fried through the intervention of sheet iron and melted grease? The nearer the fire the better the food. Take, for example, a slice of bacon, dress it in any complicated way you like, and I will engage to surpass the most intricate efforts of cookery by merely laying the bacon on the glowing and smokeless coals. It will not burn. It will curl, and coil, and twist, and splutter, as if in extremest agony; it will be lapped in fierce flames, ' like the pale martyr in his shirt of fire,' and it will pass from the flames to the table in supreme condition, without a particle of cinder upon it, with all the flavour retained, and all the superabundant grease and salt burnt out. *Expertissimo credo Roberto.*

Should any of my readers indulge in such a supper as has been described, I can predict two events but not a third. I can foretell that the supper will be a most luxurious one, and that the barrel will weigh sensibly lighter after the banquet, but I cannot predict the dreams that are likely to follow. One never knows where to stop in eating oysters. They are as insidious as walnuts or chocolate bon-bons, and the more you take the more you seem to want. 'Only just *one*' more is said over and over again, until, like the little girl in the story of the 'Three Bears,' the fascinated reveller empties the dish.

The foregoing remarks will show that the present writer is not insensible to the merits of the oyster, considered in a gastronomical point of view. Not only, however, is the oyster good to eat, but it is curious to look at, and a philosopher will not fail to afford to the mollusc a double appreciation.

See what a strange life the creature leads, fixed in some definite spot, unable to stir an inch, and enclosed between two large shelly cases. What does it eat? how does it obtain its food? and, above all, how does it convey the nourishment into its interior? Take, for example, a periwinkle, a whelk, or any similar mollusc, place it in the sea, fasten its shell firmly to some object, and in a certain time the creature will die of starvation. But place an oyster in precisely the same locality and it will thrive admirably. The secret of its life lies locked within its shells, and, if we open this two-leaved volume we shall find the whole history written within.

Granting the barrel of natives, of which we have

already spoken, let the inquiring reader exercise great self-denial, and lay aside one oyster for the purpose of examining its curious structure. We will now plunge the mollusc into boiling water for a few seconds, which will have the effect of killing it without materially injuring any of the delicate organs with which we shall be concerned. Insert the tip of the oyster-knife between the edges of the shells, force them slightly apart, and then look inside. The mass of the body will be seen in the centre, and pressed against the shell are two flat dark-edged flaps, popularly called the ' beard.' Now this so-called beard is in fact the breathing apparatus of the oyster, and has other functions besides those of respiration, as we shall presently see.

Now open the shell entirely, remove the convex valve, taking care to cut through the thick muscular attachment close to the shell, and with the point of a knife lift up these beautifully delicate membranes. They are seen to be double, like the two shells, and on tracing them round they will prove to end in the mouth, which is close to the hinge of the shell, and can be recognised by a double pair of white and pointed lips. For some unknown reason the oyster has no throat, but the mouth opens at once into the stomach, just as the outer door of a cottage opens into the sitting-room instead of the passage.

And if the curiosity of the investigator be not quite satisfied, he can easily pursue his inquiries further, and see what the oyster had for dinner, a most useful piece of knowledge to those who make their living by breeding this agreeable mollusc. Still, though we have found the mouth and ascertained the food, we have not yet discovered the method by which the food gets into the mouth.

It is found by investigation of the substances which are taken from the interior of the oyster that its food consists of the minute animal and vegetable organisms with which the water of the sea is thickly charged.

If a living oyster be placed in water, and watched while its valves are open, a continuous current is seen to run through the shells, always passing in the same direction, *i.e.* from right to left (taking the flat shell as the upper one), and running between the gill membranes. On examining the dark edges of the beard, or gill membranes, we shall find that they are divided into tiny filaments, and that each of these filaments is covered with a myriad of the minutest imaginable fibres, each of which is continually whirling with a partially spiral movement, and producing an effect to the eye as if successive waves were rolling along the surface. A similar effect may be seen when the wind rushes over a corn-field, and produces successive waves which seem to advance rapidly, though each corn-blade remains in its place.

By the united action of the countless hosts of these fibres, technically called ' cilia,' the water is forced to sweep along in one uniform direction, and, being driven between the two gill membranes, is obliged to pass over the mouth, carrying with it the invisible objects on which the oyster feeds. So powerful, indeed, is the action of these wondrous little appendages, that if a small portion of the gill be snipped off and placed in the water it will swim away as if it were living, urged by the invisible fibres, which work as briskly as ever, though severed from the body. At the mouth the lips take cognizance of the supplies, and evidently possess the power of accepting the good articles and rejecting the bad, just as the editor of a magazine decides upon the articles which daily inundate his desk.

There is yet another office performed by the gill membranes. Everyone is acquainted with that little *memoria technica* which connects oysters with the letter R, and tells us that they are out of season in the months which do not possess this delightful letter. In May, June, July, and August the oysters are not only out of legal season, but are so in literal fact, being thin, and quite unfit for food. Practically, however, the oyster season is anticipated by a month, and on August 1 the costermongers ply their busy trade through the streets; at every corner the itinerant fishmonger invites his customers to the gritty board, the great coarse molluscs, and the bottle of suspicious vinegar; while the children erect little edifices with the shells, call them grottoes, put an inch of lighted candle into them at night, and vex the souls of passengers with iterated requests for halfpence, which, by some strange logic are supposed to be applied towards the repairs of the grotto.

It would be a most injurious act to catch the oysters during these months, as they are then engaged in laying their eggs, if this strange operation can deserve the name. If, for example, a barn-door hen, instead of laying her eggs in her straw nest, were to transfer them into her lungs, there to be hatched and half fledged, we should be perplexed to find a name for the proceeding. Yet this is just what is done with the oyster. The eggs are very minute when first produced, and are kept by the parent between the shell and the gill membranes; where they remain until they are furnished with shells of their own, and able to cope with the watery world into which they are about to be launched. I well recollect, when I was a very little boy, bringing home some fresh-water mussels, and being completely astonished at finding a number of the tiniest little mussels floating in the liquid contained in the shell. So, when an oyster is out of season, and a thoughtless person ventures to eat it, he will find that a number of little shells will have an unpleasant grating effect upon his teeth, and will learn practically the effect of the ' fence ' months.

It may be said that if the female oysters were permitted to rest during the fence months, and the males brought to table, we should still ensure our present supply for the table without risking the future crops for ensuing years. But there is a difficulty here. No one knows which *are* the females. They all lay eggs after the queer fashion already mentioned, they all dismiss abundance of young oysters from their shells, and no one even knows how they do it. Hundreds of oysters have been examined by our keenest anatomists, and the only conclusion that they have decided upon is that an oyster cannot be crossed in love because there is no other sex to fall in love with, unless, Narcissus-like, the creature should suffer from disappointed affection for itself. In fact, the oyster carries out practically, with a trifling variation, the suggestion of the well-known song, and the husband and the wife can safely say that—

> They are saved so much bother,
> For they are both one another,
> And not themselves at all.

And yet the oyster is a large-hearted being, though with little brain, from which we might infer that its affections were strong and its intelligence weak, did not the previous observations prove there is no place for love. As to intellect, the creature needs but little, and has but little. It knows when to open and when to shut its shell, which articles of food to accept and which to reject, and considering the stationary life which it leads, a solitary being among thousands, like prisoners in close confinement and contiguous cells, it has quite as much intellect as it requires.

Here I find I must pause. While describing the oyster, its curious structures and habits, I recognise the same feeling which induces even abstemious men to empty a barrel of natives under the plea of ' only just one more.' The whole history and economy of this mollusc is, to me at least, so full of interest that I find myself saying, ' I will only mention just this one point,' and now discover that my paper is well-nigh at an end.   Taking leave, therefore, of the individual oyster, we will give a cursory glance to the life led by these bivalves from the egg to the table. In their very early stages we meet with the young oysters within the shells of the parent, enveloped in a gelatinous sub- stance, and partly nourished by the stream of sea-water, which washes them as it is driven along by the fringed edges of the gills. In this condition the young are called the ' spat,' and are soon dismissed from the pro- tecting shell. When set loose from the shell the young molluscs attach themselves in vast quantities to the objects on which they happen to fall, so that the nature of the bed has great influence on the perennial produce. When once fixed they increase rapidly in size, attaining the size of a florin in their first twelve- month, and are thought fit for the table when they have completed their third year. The oysters brought to market are mostly obtained by means of the dredge, which scrapes along the bed of the sea, and tears the molluscs from their attachments. This plan, however, is rather a clumsy one, involving the destruction of many young oysters, and being by no means a certain one. Efforts are now being made in many places to learn the economy of this useful bivalve, and to breed it regularly for market. During the last few years, the practical knowledge of the oyster and its habits has greatly increased, and vast artificial beds are being laid for its accommodation throughout its life. Several oyster-beds have already existed for many years, some in England, and others on the Continent. At Dieppe, for example—the only series of beds which I have examined—the oysters are managed with great care, being bred in a series of large shallow pools, and fed regularly as if they were chickens. They are all arranged in regular rows, slightly overlapping each other, like the tiles of a house-top. The green oysters, which are held in such favour, are nothing more than the ordinary species, fed for a time in ponds where the green confervoid growths are plentiful.

Without describing at length the various oyster parks which are now being established, and which, especially on the Continent, are assuming very important dimensions, a few particulars of their structure may be mentioned. Some of these parks are so extensive that they are measured by miles, and are capable of breeding many millions of oysters annually. It is found that the best substance for the reception of the spat is brushwood made into bundles, sunk under water and kept down by stones. If these fascines be removed, the young oysters are found clinging to them like grapes upon the vine, and when they are full-grown their aggregate weight is by no means trifling.

This circumstance explains the travellers' tale so long discredited, that in some places the oysters grew on trees. It is now a well-known fact that if trees growing near oyster-beds dip their branches into the water, the young molluscs are sure to settle on the immersed twigs, and by their increasing weight drag the bough still deeper. The newly-sunken branches are in their turn covered by fresh colonies, until at last the bough is fairly loaded with its strange fruit.

As far as is yet known the experiments have answered admirably, and it is sincerely to be hoped that the ingenious projectors may make their fortunes as they deserve. For it is no less meritorious to render fertile mile after mile of barren coast, to produce in countless myriads an esculent so nourishing and so palatable as the oyster, than to perform the much-lauded and laudable feat of making two blades of grass grow where only one grew before. We hope soon to sow an annual oyster crop as we now sow an annual crop of grain, for the operation bids fair to be as easy in one case as in the other, the hopes of success are equal, and the profit, if anything, rather inclines to the mollusc than to the cereal.

There are even pearls to be found in the common oyster, though they are never large enough or pure enough to be of any commercial value. I have many specimens of such pearls all procured by myself, and it is rather a curious fact that I have always found them periodically. On one occasion, out of a poor half-dozen of oysters I procured as many pearls, most of them about the size and shape of mustard seeds, but one of a pear-like form, and nearly three times as large as any of the others. We make no use of them in Europe, but in some eastern countries all the little bad-coloured pearls are burned and converted into a very delicately pure lime, for the purpose of being chewed together with the betel-nut.

I have already expressed my opinion that a connoisseur in oysters will only eat them pure and unadulterated, simply cooked in their own shells, or, more simply, without any cooking at all. But as I cannot expect all my readers to have the same refined taste, I will here present them with a receipt whereby turkey and oysters are brought into close and grateful conjunction.

Prepare the bird for boiling, open a number of oysters, varying according to the size of the turkey, and use them instead of stuffing. Put the stuffed bird into a deep jar, fill it up with milk and the juice of the oysters, and cover it over with flour paste. Boil it for four or five hours, until thoroughly cooked, serve it up very hot and with melted butter, and your guests will for ever cherish a kindly remembrance of 'Turkey and Oysters.'

DE MONSTRIS.

HAVE any of my readers made themselves acquainted with the 'Nurenberg Chronicle'? If not, they should do so. The book is very pretty reading, though its beauties are hard to see. It is a big book—an elephant of a book—a book that would have delighted the very soul of Dominie Sampson. It is printed in black letter, the language is Latin, and its subject is a chronicle of the world's history, from the time of the creation to the date of the book. It is one of the quaintest, raciest, and most delightful of books—not very easy to read, but perhaps owing some of its charm to that very fact.

It is a nut with a very hard and tough shell, but a marvellously sweet and mellow kernel. The shell is the language, or rather the mode in which that language is placed before the reader. The Latinity of the book is anything but Ciceronian, but it is couched in bold, vigorous terms, interspersed occasionally with strange words, which, on investigation, prove to be German words Latinised, the writer being unable to find any pure Latin equivalents. Punctuation seems

to have been a matter of accident rather than of intention, and the printers have thought fit to employ every
form of abbreviation that has been invented for the
bewilderment of readers. Mostly they have had the
grace to add the abbreviatory line, so indicating that a
vowel or two, or the letters 'm' and 'n,' are omitted,
and may be supplied at discretion; so that the reader
can see without much difficulty that ' $\overline{ai\hat{a}}$ ' means *animâ*.
In many cases, however, they have even dispensed with
this slight aid to the reader, and left him to conjecture
that ' *sex mans hntes* ' stands for *sex manus habentes*,
and that, throughout the book, ' *hoies* ' signifies *ho-
mines*.

The eye, however, soon becomes accustomed to these
little abbreviations (which are wonderfully like the
dress of that period—cut very short just where one
ought to expect length), and can employ itself with
the chief glory of the book, namely, its illustrations.
These are really wonderful productions of ancient art.
Wood engraving was then in its earliest infancy, and
the engraver could only produce hard, bold lines, of
nearly uniform thickness. Of perspective there was
little—of aërial perspective, none; while the art of
' cross-hatching,' so easy on metal and so difficult on
wood, had only just been discovered. Yet, notwithstanding all these difficulties, the illustrations possess,
like the letterpress, a wonderful amount of quaint,
stiff, homely power. The designs are attributed to
Wolgemuth; but, whoever may have been the artist,

there is no doubt that true art-power lay within him, and that he did as much as could be done with the very limited means at his command. They are worthy of a city which produced such men as Durer, Bebem, Hele, and Lobsinger, and are full of marvellous vigour, ' cross-hatched ' here and there, but absolutely destitute of grace, delicacy, softness, perspective, or expression. There is not a face in the whole book which has the least indication of human feeling or passion. If a murder be represented, the murderer and his victim are equally impassive of countenance ; so that if the two heads were transposed the design would suffer no injury. Facial expression was, in those days, beyond the wood-engraver's powers, and scarcely anything could be achieved except a hard, thick outline. Yet, spite of these drawbacks—any one of which seems capable of ruining an illustration—the vigorous power of the woodcuts is deserving all praise. The lines are hard and coarse, and the execution rough ; but, nevertheless, every line has its purpose, and tells its own story, much unlike the inane prettiness produced by the facile execution of our own day.

Like the letterpress, the woodcuts treat of the world's chronicle ; and by way of beginning at the beginning, the first woodcut represents the Song of the Morning Stars before the creation of the world. This rather difficult subject is represented by a circular space, quite blank, around which are tightly packed a vast multitude of heads, each crowned with a sort of tiara, and all having their mouths tightly shut, so that if they are singing at all, they must be singing through their noses. The creation of the world is ingeniously, if simply, represented, by the disappearance of the heads, and the appearance of successive concentric circles within the open space, one circle being added for each day.

The earth being made, and the fishes, birds, and beasts duly placed in it, we come to a delightful Garden of Eden, walled and castellated, and its arched entrance guarded by a portcullis, through which trickle four diverging gutters representing the four rivers of Paradise. As for Adam, he is certainly not handsome, but he is very far superior to the Adam represented in the frontispiece of a Family Bible now before me. Wolgemuth's Adam, rudely though he be drawn, is at least represented as the first man might have been; whereas the modern Adam has had his hair nicely cut, parted, and curled—has been shaved that morning, and wears a pair of neatly shaped whiskers. In fact, he looks as if he ought to wear clothes; whereas Wolgemuth's Adam looks as if clothing were no more needed by him than by a Greek statue. Then we have the creation of Eve, who is represented as being pulled bodily out of a large circular hole in Adam's side; and so we proceed with the history of the world until we come to the building of the ark.

This woodcut has a strange fascination for me, in its mingled truth and absurdity, strength and weakness.

and the under-current of earnestness that runs through the whole design. The ark is about half the size of a Thames lighter, and, small as it is, the artist has partitioned it off with scrupulous fidelity into its various compartments, and affixed to each division an explanatory label. There is an '*apotheca herborum*,' and an '*apotheca specierum*.' The clean animals are separated by the whole length of the vessel from the unclean beasts, and in the middle is the '*habitas hominum*,' about as large as a Skye terrier's kennel. All Noah's sons are dressed in the exceedingly abbreviated jerkin, the tight hose, and the pointed shoes of the period. One of them is chopping vigorously at a prostrate beam, making the splinters fly bravely; another is apparently splitting the skull of one of his brothers; while a third is seated on a raft, dubbing down with his adze the sides of the ark, which is at least a hundred yards distant from him. I have always thought that Hogarth must have seen this wood-cut before he drew his well-known 'Perspective.'

Then every place that is mentioned, whether it be city, country, or continent, is illustrated by an engraving, more remarkable for its ingenuity than its accuracy. Why '*Provincia Britannia*' should be represented by thirty or forty castellated towers, crushed together within a circular wall, and let neatly into the sea, is a mystery not easily to be penetrated. Supplying Nineveh with a cathedral inside the walls, and a quintain outside them, is a venial error in chronology, as are the soldiers in plate armour by whom Abraham is accompanied, and the clock in the parish church of Damascus.

Then there is a charming audacity in the way in which the same woodcut does duty for several subjects. Take, for example, the two woodcuts which have just been mentioned. Nineveh, which is figured in page 20, becomes Corinth in page 33. Then the same cut which represents Damascus at an early period of the History, becomes successively, Macedonia, Verona, Ferraria, and the Province of Germany. The portraits are just as versatile. There is one, for example, which so irritated me by its persistent reappearance, that I took the trouble to hunt it through the book. It represents a close-shaven, wizen-faced, vulture-nosed, under-jawed old gentleman, wearing a kind of fez cap, and reading a book without looking at it. On page 59 he is Solon; on page 80 he is Demetrius, and only two pages further he is metamorphosed into Paretius. On page 111 he is Suetonius, and on page 158 the Venerable Bede. On page 200 he is St. Hugo (*natione Gallus*); on page 213 he does duty for Barnardus of Compostello, doctor; and on the very next page he is labelled as Alexander, *doctor irrefragabilis.* Page 227 reveals him as 'Johannes de Monte Villa, *eq. aur nat anglic*;' and we finally take leave of him on page 240 as 'Johannes Gherson, *cancellarius Parisiensis.*'

I have already mentioned that facial expression is absolutely wanting in these various portraits, so that, judging by the expression of the countenance, no one can tell whether the face is that of a hero or coward, an honest man or a thief, a saint or a scoundrel. The artist, however, is at no loss for means by which to indicate the moral condition of his subjects, especially those of the female sex. Drapery invariably represents piety, and moral excellence may be measured by the amount of folds in which the various personages are enveloped. Sanctity is always clad in flowing robes, while a curtal frock is a sure sign of a guilty conscience. Lot's wife, Mrs. Potiphar, and other doubtful or objectionable characters, are clothed in short and scanty raiment; while all the female saints (with the exception of Mary Magdalene, who wears nothing but her own hair, and plenty of it) are endowed with trailing skirts, proportionate in length and volume to the sanctity of the wearer. The culmination of drapery, however, is to be seen in the wonderful print representing the Adoration of the Magi. In this print the most conspicuous object is the drapery with which the principal figure is clad. Her mantle is large enough for a tent, and all in waves, like a sea, over the foreground. It is carefully gathered into a thousand angular folds, as if made of the stiffest fabric, and has more than enough material to clothe fully all the scantily-dressed sinners in the book.

As may be imagined from the character of the book, both author and artist revel in the various monsters which were fully believed to inhabit the world. Two whole pages are given up to these crea-

tions of the brain, and very odd beings they are. Some
are supplied with explanatory descriptions, but others
are left to the discrimination of the reader. Enormous development of some part of the human body is
the usual method in which the various monsters assert
themselves. There are the Pannothi of Scythia, whose
ears are so large that they cover the shoulders. There
are men with enormous under lips, which fall over
their breasts. There are the Unipeds, men with only
one leg, but then the foot makes ample amends for the
absence of a second limb. It is about as large as an
ordinary umbrella, and is used for the same purpose—
the Unipeds being in the habit of lying on their backs,
and sheltering themselves from sun or rain by the
enormous foot. These men can of course only hop,
but they do so with such prodigious celerity that they
chase and catch stags by hopping after them.

Sometimes the monsters enjoy a superfluity of
members. There are men with two heads, like the
Welsh giant who was ignominiously worsted by Tom
Thumb. There are men with four eyes, men with four
arms, and men with four legs, balanced by Monoculi,
or men who have only one eye—which is set in
the middle of the forehead—and men who wear no
particular head, but have their eyes, noses, and mouths
placed in their breasts.

Of course there are dwarfs, and a single combat
between a pigmy and a crane is drawn with some
vigour. The crane seems likely to get the better of the

pigmy, for his beak is long enough to spit the little man through and through. Then come the monsters of the composite order, *i.e.*, which are made up of beast, bird, and man. There is a very fine Centaur; there is a man with a human head perched at the end of an ostrich's neck; and there is a man whose head is adorned with a large pair of ibex horns—'*quales in solitudine S. Antonius Abbas vidit.*' Then there is a nation of women who have beards flowing over their breasts; and there is a whole nation of Hermaphrodites, of which remarkable beings the artist gives an authentic portrait. The portrait represents a human being, man on the right side and woman on the left, divided accurately by a perpendicular line down the middle of the body, much like the well-known portrait of the Chevalier D'Eon, or the older and wider known print of 'Death and the Lady.' The head has a most absurd appearance, being furnished with short hair and half a long beard on the right side, while the left has a smooth face and long straight hair. And after being gratified by the sight of these wonderful beings, the reader is confidingly told that there are yet many more monsters in the world—'*quos commemorare perlongum est.*' There is an evident good faith in all these queer drawings, and the equally queer descriptions; and it is quite delightful to read a book in which these absurdities are gravely treated as true.

On examination of the series of monsters, there is

one point which can hardly fail to strike anyone who has some smattering of physiology, and has taken some interest in ethnology. They are all very odd, not to say ludicrous; but there is very little invention about them. Nothing is simpler than the notion of enlarging, multiplying, or curtailing the various members of the body; so that the headless men, the two-headed men, the one-eyed men, and many-eyed men, the six-handed men, the big-footed men, the long-necked men, the large-eared men, and the flap-lipped men, are all off-shoots of a single idea. Then all the composite monsters are but off-shoots of the one single idea of grafting parts of other animals upon the human form, and there is really but little invention required in carrying it out. You may boldly join two bodies together, as the Centaur, the harpy, and the mermaid; or you may tack parts of one body to another, as the satyr, with his goat's legs, the faun, with his pointed ears and little tuft of a tail, or the man with ibex horns on his head. Again, several of these monsters are nothing but exaggerations of actual fact. It has long been suspected that the ancients knew more of the world than has generally been supposed, and this idea is strengthened by ex-amining the monsters of the Nurenberg Chronicle. Take, for example, the men with the enormous under-lip. It is highly probable that the idea was taken from the narration of some traveller, who had seen one of tne many savage tribes which distend their lips, either upper or lower, by the insertion of circular pieces of wood in them. The long-eared men can be accounted for in a similar manner. All ethnologists know that there are many tribes which measure their gentility by the length of their ears; and by cutting holes in their lobes, and hanging weights to them, succeed at last in getting them to hang down on the shoulders, just as is represented in the Pannothi. As to the nation of pigmies, we all know of several tribes or nations that may very fairly be called by that name; and although they are not so very small as is represented in the woodcut, they are yet so small that when standing by the side of an European of middle stature the grown men and women seem scarcely bigger than our children of nine or ten. The many-armed monster is evidently borrowed from the Indian temples, which are covered with statues of various deities, scarcely any of which condescend to have less than eight arms. The two-headed monster is due to the same source.

There is another point which is worthy of notice. Putting aside the centaur, harpy, and mermaid, there is scarcely any of these monsters which is not a real fact. Take, for example, the horned man. Such beings really have existed, as is well known to physiologists. There is now in the museum at Oxford a portrait of a woman who was remarkable for possessing several horns on her head, and by the portrait is deposited one of the horns in question. After all, there is nothing so very much out of the way in this curious development. For all the hollow horns, such as were these, are nothing but modifications of hair, and it is not an uncommon phenomenon to find a different mode of development of the same material. In the museum of the late and regretted Mr. Waterton there is the head of a sheep, with a horn growing at the end of its ear, and not on its forehead—the hair having been modified into horn, just as was the case with the horned woman. The feathers of birds frequently undergo a similar modification; and there is a well-known instance of a whole family of human beings—popularly called the 'Porcupine Family'—many of whom had the whole body covered with small horny growths.

Then the two-headed man is evidently a reminiscence of one of the many known instances where twins have been partially fused together. In the present day we have the long-known 'Siamese Twins,' the 'Two-headed Nightingale,' and the 'Ohio Twins,' the latter being united from the back of the head down the entire spine. Plenty of similar instances have been known. There were, for example, the 'Biddenden Maids,' whose memory is still cherished in the neighbourhood where they lived; and I have before me a collection of drawings illustrative of similar strange developments.

So with various other forms. The headless man only represents a fact well known to every physiologist; while the very bold stroke of the artist in putting the eyes, nose, and mouth into the breast, is not without its parallel in nature. Multiplication or curtailment of limbs is very common, and often runs in families;

so that the artist of the Nurenberg Chronicle really needed no great power of invention in many of his representations. As for 'women with beards descending on their breasts,' there is nothing very remarkable about them, and plenty of instances have been known. Perhaps some of my readers may have seen the late ' bearded lady,' Julia Pastrana, whose preternatural ugliness, quick intelligence, proficiency in modern languages, and agility in dancing, were, some years ago, quite familiar to the British public. Her beard, though it did not quite descend to her breast, was stiffer, thicker, and longer than that of many a man ; and yet her hair, a lock of which lies before me, was not coarser than that of an ordinary Spanish woman. And there was a lady, very well known in a certain cathedral city a few years ago, who possessed a moustache as thick and full as that of a life-guardsman, and a beard of very fair dimensions.

Bearded women naturally lead us to the hermaphrodites above-mentioned. I fancy that the idea was originally taken from travellers' accounts of certain divisions of the Malay race, in which the two sexes can scarcely be distinguished from each other, except by the fact that the women look much more masculine than the men. Naturalists all know that, although among the human race such a man-woman has never been known, there are many creatures in which such a phenomenon does take place. There is, for example, scarcely any large collection of insects, whether public

or private, which does not contain specimens of insects
—generally butterflies—which are male on one side
and female on the other: the division being along the
centre of the body, exactly as is represented in the engraving of the Nurenberg Chronicle.

Indeed, in actual nature are to be found almost
every instance of monstrosity which the mediæval mind
could invent, and plenty besides, which would throw
these mediæval monsters entirely into the shade. For
example, the Monoculi, or one-eyed men, are represented by a wonderful number of the tiny crustacea
called *Entomostraca*, many of which are remarkable
for having a single eye placed in the midst of that
part of the body which does duty for a head, and which
are in consequence called by the appropriate names of
Cyclops, Polyphemus, and so on. As to the many-
eyed men, they are entirely outdone by the insects,
some of which have more than thirty thousand eyes—a
fact which infinitely exceeds the imagination of the old
writers or artists. The men with their faces in their
breasts are represented by the crabs, lobsters, shrimps,
and their kin, all of which have their mouths set
exactly in the situation in which the old artist has
placed the mouth of his headless man.

Then the unipeds are far outdone by many of the
molluscs, whose structure may be characterised as a large
foot, carrying a comparatively small body on it. In
the Nurenberg Chronicle the uniped is figured as
using his foot as an umbrella; but neither artist nor

author would have dared to describe him as using it in the light of a boat. Yet this is actually done by many river snails, whole fleets of which may be seen floating leisurely down the stream, the large foot being hollowed along the centre, and acting as a boat. There is no need to multiply instances, but we may receive it as an axiom that there exist in nature monsters far more wonderful than any exaggeration or combination that has been invented by man.

I cannot leave the subject of monsters without a short reference to the Japanese 'mermaids,' which are now and then brought before the public. These are nearly all made to represent the conventional idea of mermaids, except that the upper half is formed in semblance of a monkey, and not of a human being. They are quite common, and are manufactured by dozens, most of the makers adhering to the same type, only one or two striking out original ideas of their own. I have seen one specimen in which the maker had the audacity to add a pair of large wings formed like those of the bat.

They are well made, but not so well as is generally thought. The late Mr. Waterton, whose skill in taxidermy was supreme, had an entire contempt for Japanese mermaids, which he stigmatised as clumsy fabrications, saying that he could make better work with his left hand. Certainly the amusing monstrosities which he made, and with which he delighted to delude visitors to his collection, were much superior to the best Japanese mermaid that I have seen. Some years ago, a fishmonger in the old Hungerford market showed me one of these mermaids, and was quite angry with me when I praised the excellence of its manufacture. He really believed that it was a genuine inhabitant of the water, framing his belief on the fact that there was no junction between the fish and the ' maid.' Neither was there. The Japanese taxidermist knew his business too well to have any junction at all, the seeming skin being nothing but *papier maché*, worked over a model, and having fins, scales, teeth, and nails inserted in the proper places.

OUR RIVER HARVESTS.[1]

MAN can endure many things. Incredible as the assertion may appear, civilised man is capable of maintaining existence, though deprived of a town and country house, a box at the opera, three or four gigantic footmen, and a velvet-footed valet. Follow him downwards through all the phases of terrestrial conditions, and under all varieties of climate, and you shall find him by degrees casting off one garment after another, and one want after another, until the primitive savage stands before you, wholly without clothing, and almost without wants.

He needs no tailor to shelter him from the cold, for his body is 'all face;' and in particularly severe weather he clothes himself by the simple process of stripping the skin off some newly-slain animal and flinging it over his shoulders. He needs no architect nor builder, no carpenter and no plumber to aid him in erecting his house; for the cleft of a rock, or a hole

[1] The following article, which was written in 1862, is here given in order to show the science of Fish-hatching as it was in its earliest infancy. Many improvements have been made, and several of the abuses which are mentioned have been corrected.

rudely scraped in a bank, is all the home that his imagination can conceive or his needs require.

But however widely different may be the polished exquisite of London society, and the rude savage of the antipodes, they both agree in one absolute want—

HATCHING-TROUGHS IN GREENHOUSES.

namely, that they must eat or die; and if we trace effects to their causes, we shall find that our lamented friend Soyer was not very far wrong in considering the culinary art to be the mother of civilisation.

As with individuals, so with nations, which, after

all, are but the aggregates of individuals.   As a general
rule, a hungry man becomes uncivilised in proportion
to his hunger, and his diminished powers of argument
are proverbial.   Even the compulsory postponement of
dinner for an hour or two has a mightily injurious effect
on the best-tempered of Britons; and it may easily be
imagined that in the primitive ages of society, where
no one ever has any dinner, and is always waiting for
his food, the general character will be wolfish, snappish,
tetchy, and selfish.   Food and civilisation are connected
together by indissoluble bonds, and the first necessity
of a civilised country is that its food shall be plentiful
in quantity, good in quality, and readily procurable, so
as to insure the periodical recurrence of nutritious
meals.

There can be no true civilisation where every man has
to hunt for, kill, carry home, and cook his meals, inasmuch as he thereby lowers himself to the grade of a
mere animal of chase, and is forced to give up all his
finer faculties to the one task of seeking and eating his
prey.

The prosperity of every country depends chiefly
upon its supply of food, and a nation advances and re-
cedes exactly in proportion to the quantity and cheap-
ness of provisions.

It is true that the present state of commerce enables
nations to interchange their commodities, and to supply
the non-harvesting lands with the food which they need
but cannot raise.   But there are many turns in the

wheel of fortune which may cut off the supply, and which might deprive one nation of food just as another nation is deprived of cotton. Beyond all value, therefore, is the power of being independent with regard to food; and the nation which can discover a fresh indigenous supply has made no small step in her prosperity.

We have long ago realised the value of the land as a food producer. The ancient forests are falling before the enclosure acts, like grass before the mower's scythe; the wide commons are gradually changing into meadows and farm-yards; and their gold-blossomed furze bushes and purple heather are forced to make way for the less picturesque, but more valuable, corn and turnips; and even the very banks of railway cuttings are economised by thrifty workmen, and yield their crops to the strong hand and skilful brain. Chemical agriculture has now advanced to the rank of an acknowledged science, and the most unpromising soil is rendered fertile by the judicious addition of certain elements which the desired crop demands, but which the ground does not possess.

We have partially realised the value of the sea which surrounds our island, and have learned that the edible treasures of the ocean are priceless as inexhaustible. Let the fisheries of herring, sprat, flat fish, cod, mackerel, and pilchard be destroyed, and the shock to English prosperity would be severe as the shattered credit of a bank to a trader who has entrusted a great part of his capital to its keeping.

We have still, however, another source of national greatness—a very gold mine of wealth—requiring little outlay and less trouble. Our rivers bring riches to our very feet ; and the golden sands of the Pactolus may be outshone in true value by the pebbly gravel, stony rocks, or shadowy banks of our English streams. The treasures of California and Australia lay hidden in the rocks and sands for ages, trodden under foot by the heedless and ignorant, and only revealing themselves to those who would work and think. In like manner the treasures of our own streams sweep daily past our unsuspecting eyes, and will be given only to those who will take the trouble to learn about them and search for them.

It is but lately that we have begun to think that good fish are really valuable articles, and to discover that the supply is annually decreasing. For this discovery we are chiefly indebted to sportsmen, whose observant habits and watchful acuteness are invaluable aids to the cause of which we are about to treat. And, although in the few pages which can be given to an important subject we shall treat of the rivers and their living treasures without any reference to mere sporting interests, the reader will of course understand that the interests of the nation and the sportsman are identical, and that, in speaking of the one, we necessarily include the other.

For the last few years our river fisheries have been failing. There is no doubt of it. The sporting papers are full of complaints respecting the decrease in size and quality of the river fish, those of the salmon tribe being most conspicuous in this respect. The complaints have waxed louder and more frequent, until they have ceased to belong to the mere sportsman, have found their way into the general press, and become a question belonging to the nation at large. Even the leviathan of the press has more than once taken up the subject, and drawn the public attention to the dismal fact.

It is no light matter that all the best fish should be gradually extirpated from our rivers ; and the decreasing numbers of the salmon alone afford grounds for just fears. The salmon ought to be as cheap as the herring, if not actually cheaper. It needs not to be brought from the sea at a vast expense of fishing-boats, nets, and all kinds of auxiliary apparatus, but, if properly managed, will bring itself to many an inland town, and need only the trouble of catching.

There never was a more obliging creature than the salmon. It will provide for itself entirely. It wants no shepherd, no hurdles, no folds, and no food. Its habits are regular as clockwork. Given the young salmon, and you will know exactly where he is, and what he is doing, on any day of the year. He grows out of his baby-clothes in the river, never stirring very far from his cradle ; and then somewhere about his second May he puts on his first suit of silvery scales, and makes for the ocean. He remains in the sea for a certain period, feeding vora-ciously on the rich banquet which the salt waters produce; he gets himself into admirable condition, becomes as fat as a pig and as firm as a turkey; and when he is quite fit to be eaten, back he swims to his native river, and comes to be killed with the proverbial docility of Mrs. Dilley's ducks, weighing as many pounds on his return as he weighed ounces on his departure.

His flesh is very digestible, wonderfully nutritious, more so indeed than that of most fish, and its only fault is its luscious richness. ▸It can be eaten fresh, pickled, or dried; and in the last-mentioned case can be preserved for years in perfectly good condition. And, as the salmon feeds himself, the cost of his maintenance is *nil*, and the only expenses connected with the culture of this noble fish are the salaries which must be paid to the water police.

The salmon ought never to have occupied the position which it now holds, namely, a dainty upon the tables of the affluent. The poor man ought to have his salmon as well as the rich; and if the newly-born science of pisciculture should prosper, a few years will see the labourer or the mechanic purchasing his salmon as freely as he now purchases his herring or periwinkles. There was once a time when this splendid fish was so plentiful in the British rivers that apprentices were accustomed to stipulate with their masters not to be fed on salmon more than four days in the week; and though we cannot hope to restore the fish in such plenty as is indicated by that arrangement, there is every hope of bringing them back to the rivers which they have deserted, and retaining them in those out of which they are now rapidly disappearing.

The first step is evidently to find out the causes which drive them out of our streams, and try to rectify them ; for it is clear that to stock a river with salmon would be an useless expenditure of time and trouble, when every fish is sure to be killed before it is as big as a sprat.

Perhaps there is no fish so systematically persecuted as the salmon; and when all the circumstances are reviewed, it is really a wonder that a single salmon ever attains its full size. From the time when the unhatched egg is deposited in the river, to the time when the fish returns to fulfil its great office, every yard of water contains a foe, and every mile of river conceals a trap. The eggs are surreptitiously taken by poaching anglers, and used as bait for other fishes, and the young fry when hatched, and just able to move, are gobbled up in thousands by various finny depredators and water fowl of different kinds.

But, putting aside these natural enemies, which, after all, only preserve the due balance of nature, the artificial impediments with which the fish meet are numerous and fatal to a degree. These impediments may be divided into two classes—the fixed and the moveable. The latter term includes spears, or leisters— terrible instruments, like Neptune's trident, on whose barbed prongs the salmon is impaled as it lies on the bed of the river—and nets of all kinds, ingeniously made so as to sweep the whole breadth of the stream, and to entangle even fishes of a few inches in length.

As to the fixed impediments, their name and structure is legion. 'Weirs,' or barriers, are made of timber, or even faggots, and so constructed as to intercept almost every fish as it tries to make its way along the stream. 'A Devonshire faggot weir,' writes a correspondent of the 'Field' newspaper, 'for thorough impassability, in some ninety-five days out of every hundred, almost baffles description: extending the whole breadth of the river, staked, ruddled, stumped, and twisted, leaving out long bushy ends down stream, partially filled up with large stones, often some ten or fifteen feet wide at the top—is so admirably constructed for stopping even a minnow, that the whole stream drains and percolates through this mass of bushes. In many places a London lady could, with little trouble, walk over dry shod.' The same writer, after dilating on the many impediments placed in the way of these migratory fish, proceeds to remark that in hot weather, and after a dry spring, the young salmon perish in vast quantities while trying to force their way through the mazes of the brushwood, and taint the air around with their decaying bodies.

Water-mills are notoriously employed for the illegal capture and destruction of the salmon, both in its early stages and during the fence months; and the destruction of 'foul' fish, as they are then called, is almost beyond belief.

It is true that the fish when foul acquire a peculiarly unpleasant taste, the flesh loses all its firmness, becomes loose and flabby, and gives forth a very unpleasant odour. A Scotch peasant will have nothing to do with the foul fish, as far as eating it goes ; but the French have an idea that it is a great dainty, and accordingly pay high prices for the worthless article. The natural consequence is that many tons weight of foul salmon are illegally captured and sent to France, where they appear on the tables of the *bons vivants*, and are lauded to the skies by the guests, their chief value consisting in the fact that they are brought all the way over the sea, and cost much money.

It is true that a penalty is attached to the act of killing foul or unseasonable fish, and now that an Act on the subject has passed through Parliament, the fish may stand a better chance of attaining their full growth. But it is useless to affix a small or even a moderate penalty to the transgression of this law, as its infraction is so profitable that the offender will compound for a dozen detections, provided that he succeeds once in capturing and selling the illegal booty. If a man can make between three and four hundred pounds by one capture, he cares little for a few fines of a pound or two each.

Mutual jealousies of neighbouring proprietors cause the destruction of young salmon in vast quantities, each owner being anxious to secure the fish while he has the chance of doing so, and being unwilling to allow his neighbour the benefit of the fish which pass through his waters. Each goes upon the argument that if he, Mr. A., does not catch the young salmon, Mr. B. or Mr. C. will be sure to do so, and as the little fish are better eating than trout, he may as well obtain the benefit of them while he can. It will be seen, therefore, that supposing the stream to belong to twenty proprietors, and that nineteen of them agree to permit the salmon a free passage through their domains, the recalcitrant twentieth may neutralise all their efforts, and, by fixing weirs and using nets, may intercept every fish as it passes.

There is another reason why the salmon is driven away from many rivers in which it was formerly plentiful, namely, the polluted state of the water. The Thames enjoys an unenviable pre-eminence in foulness; and though the last two or three years have seen a slight improvement in its condition, it is quite impossible to predict that a recurrence of the pestilential odours of 1857–8 may not happen in any summer. Even the famed waters of Marseilles harbour, which are said to be extremely valuable to mariners because they kill even the barnacles that adhere to ships, and the molluscs that bore into timber, can hardly be more detestable than the once silver Thames on one of its bad days. No fish can be expected to live under such horrid conditions; and though a salmon will endure much in its instinctive desire to ascend the stream, it cannot pass through such a horrible element in safety.

There are, it is true, some rivers where the water is quite as nauseous as that of the Thames, which are yet ascended by the salmon. But in these cases the extent of foul water is comparatively small, and the fish is enabled to pass through it at a single run, whereas the length of polluted water in the Thames is so great that the salmon would be forced to endure a day or two of foul water before it could gain the comparatively sweet waters of the upper river.

Recent improvements in the drainage may, however, have a beneficial effect upon the Thames; and if the waters can be rendered sweet enough for the salmon to live in, and kept clear of nets and weirs, we may look forward with some hope of success to the reappearance of this noble fish in our noble river.

Now let us glance at the means by which it is hoped to restore the salmon and other fish to the waters whence they have been extirpated.

The early life of all fish is most precarious, and from the time that the eggs are first deposited in the river to the time when the little creature is sufficiently strong of fin to take care of itself, a host of enemies surround it, and its chance of life is scarcely more promising than that of a tender little lamb among a flock of wolves. What with creatures that eat the spawn, creatures that devour the fry, and infernal engines that destroy the growing fish, not one-hundredth part attain even to their white smolt robes, and not a thousandth part reach maturity.

Many persons, even those who have taken a personal interest in the advancement of this new science, have an idea that the object of a pisciculturist is similar to that of a game preserver, namely, to furnish anglers with sport in rivers whence the fish had been driven, or in which they had never taken up their abode. It will, however, be shown, in the course of the following pages, that the question is one of national importance, involving the supply of food to the masses, and not intended only to furnish amusement to the few.

The first point in the rearing of fish is evidently to shield the eggs and fry from all their preliminary danger, and to keep them in some place of safety until they are strong enough to take care of themselves. The only method of accomplishing this purpose is evidently that the place where the little creatures pass through their first stages of development shall be either wholly separated from the river, or so carefully fenced off by close wires, that the predatory fish and other foes shall not be able to gain admission.

Several modes of isolation have been invented and worked with success; and the public are already familiar with the names of Stormontfield and other places where the breeding of fish has been tried. It is not needful, however, to go so far from home for such experiments, as an establishment which has lately been mentioned in many of the daily papers is in full operation near the banks of the Thames, under the superintendence of S. Ponder, Esq., of Hampton, who has erected and still maintains the greater part at his own expense. This, although on a small scale, is marvellously successful, and is capable of producing nearly one hundred thousand fish annually.

The process is as follows :—

Within a moderately-sized green-house have been erected a series of troughs formed of slate, and arranged one above the other like so many stairs. Each trough is three feet in length, seven inches in width, and five in depth. It is found, however, that the depth need not exceed four inches. In these troughs is placed a layer of moderately fine gravel, about two inches in depth, and larger stones are stuck into the gravel at intervals of an inch or two. The gravel and stones have been previously boiled and washed, in order to destroy all traces of decaying animal matter which might taint the water, all aquatic creatures which might injure the eggs or fry, and all confervoid growths which might choke up the stream and interfere with the wellbeing of the young fishes.

Above these troughs is placed a large tank holding about two hundred gallons of water, which is conducted to the upper trough by means of a pipe and stopcock. At alternate ends of each trough is placed a short pipe, which conveys the water from one to the other, and in consequence of their alternate arrangement compels the water to traverse the entire surface of the gravel.

The eggs are carefully laid upon the gravel so as to lodge in its interstices, each trough containing three

T

thousand ova.   As, therefore, the percentage of unhatched eggs is extremely trifling when they are in
proper condition, this single set of troughs can turn out
about thirty thousand young fish at a single hatching.

HATCHING-PLATE.

An experiment has been successfully tried to substitute slabs of slate for the gravel, the plates being
exactly one foot long and seven inches wide, so that
three will precisely fit into each trough.   The plates

EGG, FRY, AND PARR.

are covered with cup-like hollows, much resembling the
little pits in a ' solitaire ' board : a small hole is pierced
quite through the centre of each, so as to permit water
to pass freely.   Each plate contains one thousand of

these cups, and each is intended to hold one egg, so that the tedious process of counting the ova is no longer required.

When the eggs are properly arranged the water is permitted to flow very gently over them, and its force is gradually increased until it imitates as nearly as possible the shallow rippling part of the stream where the fish generally lays its eggs, and the motion of which seems to be essential to the hatching of the egg. The stream is about one inch in depth.

One great advantage of this plan is that the eggs and young are always kept in view, and are at a convenient height from the ground, so that they can be watched with a lens through the crystalline water, and their changes noted from day to day.

Another establishment is placed in the open air, not very far from the banks of the Thames. This consists of a series of flat troughs made of elm, and measuring four feet in length, fifteen inches in width, and eight in depth. These troughs or boxes are furnished with gravel and stones, as has already been mentioned; they are set end to end, and water flows continually through them from a little spring which has been ingeniously diverted in the proper direction.

The eggs are placed in the upper boxes, covered with coarse gravel, and the water suffered to flow gently over them until they are hatched, an event which usually takes place in sixty or seventy days. The temperature of the water has, however, much to do

with the time occupied in hatching. In this establishment, where the water is kept at a tolerably uniform
temperature of 45° Fahr., the commencement of the
process is visible in fifty-five days, the action of the heart

OPEN-AIR TROUGHS.

being perceptible even to the naked eye, and a singularly beautiful object under the microscope.

When first hatched the young fish is a most
curious little object, having a thin, long, transparent
body, hardly visible when immersed, and bearing the

yolk of the egg attached to its abdomen. This egg vesicle is of a bright reddish orange colour, traversed by the minutest imaginable vessels of bright scarlet, and remains visible for about seven weeks. As long as the little creature retains this vesicle it needs no food, and takes no trouble about feeding until it has lived for seven or eight weeks, when the supporting vesicle is absorbed into the body, and the fish is then thrown on its own energies for subsistence.

Now comes a critical time in the life of a fish, and one where many pisciculturists have failed. What is the little creature to eat, and how is it to obtain its food? Liver, dried and reduced to powder, is held in some estimation, and so are little worms and caddis chopped very fine. Stale bread grated into a fine powder is another useful kind of food. But it often happens that the little fishes are so delighted with the unaccustomed gratification that they continue to gorge themselves until they die of very repletion. Sometimes the pisciculturist who has succeeded in hatching the eggs under cover forgets that his little favourites must needs eat, furnishes them with no food, and so lets them perish of slow starvation after the supporting vesicle has been completely absorbed.

Practically it has been found that the combination of the slate troughs within the house, and the wooden boxes in the open air, afford the best chance of success, the young fish being removed from the former to the latter after they have assumed the perfect shape and begin to feel the want of food. This food they will then find for themselves. The sharp-eyed little creatures can see and capture the myriad tiny inhabitants of the water which are too minute to be detected by the human eye, and when they get a little stronger may be seen jumping at midges with wonderful boldness and activity.

Their peculiar habits are a sufficient guide to their owner as to the time when they are fit to enter the river and be turned loose on existence, for they drop down from box to box according to their development, and those that are found in the lowest box are always strong enough to be removed.

It should be mentioned that gratings of perforated zinc form an impassable obstacle to the escape of the young fish or the entrance of obnoxious intruders, while covers of the same substance are fastened over them at night, and are replaced in the daytime by strong netting. The object of this change is twofold : firstly, that the midges and other little insects on which the fish feed should have free admission to the surface of the water ; and, secondly, that the inmates should be guarded from various predatory birds, kingfishers especially, who would hold high revel over so plentiful and delicate a banquet.

When the fry have attained a moderate size they are removed from the lowest box, placed in a proper water vessel, and transferred to a boat. The owner rows gently about the river, and wherever he sees a favourable-looking spot he puts a hundred or so young fish into the water. No sooner are they in the river than they act as freely and boldly as if they had passed all their little lives in its stream, dive down at once, and ensconce themselves among the pebbles. This instinct is most valuable, as the fish know by its wondrous power how to separate from each other, and take up their abode in little nooks and crannies, where not even the voracious perch can get at them—at all events, not without an expenditure of labour which that fish is not at all likely to employ.

The plan of putting them into the river in little detachments is followed because it is found that whenever the little fish are thrown into the water wholesale, the larger river fish make a great feast on their little visitors, charge fiercely at the crowd, and more than decimate their ranks before they can conceal themselves, their very numbers preventing them from finding the shelter which their instinct urges them to seek.

They are very pretty, these little fish, and even in their very young days possess sufficient individuality to mark each species. The young salmon fry, for example, are rather long and slender in proportion to their width, and their hue is ruddy brown, barred with dark patches on the sides. The young trout are shorter, thick and dark, and the barred surface is perfectly conspicuous even when the little creatures do not measure one inch in length. The char are light grey above and silvery white beneath, and have a peculiar darting action, flashing through the water like a miniature rocket, and just turning on the side so as to suffer a silvery gleam to appear for a moment and then vanish.

The eggs are delicate globular bodies, varying in size according to the species of fish from which they come. Those of the salmon are about the size of sweet peas, and the perfect, healthy, and vivified egg has a peculiar translucency, with pink or ruddy reflections as the light passes through its substance.

As the eggs approach maturity the blood-red hue deepens, and when the little fish makes its escape from the imprisoning envelope the egg vesicle retains its warm hue. Indeed, it is only by this vesicle that the presence of the very young fish can be detected as they lie among the stones, the delicate bodies being of such glassy transparency that they would escape observation but for the ruddy hue of the egg vesicle which is attached to them.

Should the egg be unfortunate, and its vital principle escape, the fatal result may be at once known by the change of aspect, a gradually increasing opacity spreading through the whole substance and the egg looking exactly like a boy's 'alley taw' seen through the wrong end of a telescope. By degrees a kind of flocculent excrescence begins to grow upon the egg, and it is soon surrounded with this growth to such an extent that it becomes as large as a moderately-sized gooseberry. All such eggs must be removed from the water, or they would otherwise taint its purity; and as the increased bulk renders them lighter than the element in which they lie, they float to the surface and are readily detected.

Whether the eggs are hatched sooner in the artificially made gravel beds of the troughs than in the natural gravel of the river is not very clear, but it is certain that even in the open-air boxes, where all conditions are apparently identical, the salmon eggs are hatched in little more than half the time which the generality of books mention as necessary for that operation. It is hardly needful to say that the rapidity of hatching is an important element in pisciculture, and that the breeding apparatus is rendered more valuable in proportion to the number of hatchings of different fish it can turn out in a season. After each hatching it is as well to remove the gravel, wash the troughs thoroughly, and not to replace the stones and gravel until they have again been submitted to the ordeal of boiling water.

The question of mixed or hybrid breeds is now attracting considerable attention, and many thoughtful inquirers are endeavouring to produce mixed breeds of fish just as enterprising agriculturists produce breeds of cattle. It seems to have been tolerably well proved that with trout the surest method of obtaining the heaviest and finest fish is to introduce continual additions of new blood into the establishment, so that the dwindling process which generally happens when the 'in and in' system is adopted may be obviated, and a fine and healthy offspring be the result.

Some experimentalists have mooted another question, namely, the possibility of crossing the salmon with some other fish, so that the offspring may retain the size, flavour, and beauty of the salmon, while the migrating instinct may be eradicated. I do not think, however, that any such attempt can be successful. In all the history of cross breeding the results prove that it is always possible to introduce an instinct, but that to eradicate one is a task almost if not quite impossible. The outward form may be alterable to any extent, but the inward character will remain.

In the greyhound, for example, when the breed was found to gain speed at the expense of courage and endurance, relinquishing the quarry at the first check, a cross of the bulldog was introduced. In a few generations the clumsy head and short limbs of the bulldog were eradicated, but the indomitable courage and tireless perseverance have remained, and the result is the present breed of greyhounds, which will not only run like the wind, but are marvellously enduring, and when they have once been set on the track will continue the chase until they drop from fatigue, or even die on the spot. Taking these and other similar examples into consideration, I cannot but think that the result of crossing the migratory salmon with some stationary species would have precisely the opposite effect to the intentions of the pisciculturist, and that, instead of making the migrator stay at home, the cross would only send the non-migrator off to sea.

Moreover, to obtain a hybrid by means of crossing two distinct species of fish is a very different business from getting a mixed breed of varieties belonging to the same species of cattle. And although it is true that even in the wide seas specimens are now and then caught which possess the characteristics of two separate species in such equal proportions that they cannot be referred with certainty to either, yet these exceptional cases prove little but a fact already known; and though they show that hybrid or mule fishes can be obtained, they fail to demonstrate any advantages to be gained by them.

We have said nothing as yet with respect to the means by which the eggs are obtained by the pisciculturist. It is, of course, necessary to be perfectly sure of their genuineness, and the only method by which this question can be decided with absolute certainty is to procure them from the parent fish.

Nothing is simpler than this process. At the spawning time, just when she is about to deposit her eggs, the female fish is put into a tub with water, and by a little artificial aid the whole of the eggs, or ' hard roe,' are soon laid in the tub. The fish is then released, and suffered to return to her native river. A male fish of the same species is then put into the same vessel, and some of the milt, or ' soft roe,' is deposited in a similar manner. He is then set at liberty, and the water stirred about for a few minutes, when it becomes cloudy, as if milk had been poured in it, but soon regains its former clearness. The eggs are then rinsed with fresh water, and are fit to be put into the trough.

Indeed, the whole process of hatching the fish is so simple and easy that it may be achieved with a flower-pot and a watering-can, and conducted on a drawing-room table. Anyone can do it, and it is really so elegant and interesting a process that it may possibly become as fashionable as the ferneries and aquaria of the present day.

Vivified eggs can now be readily procured from many parts of England and some portions of the Continent. For the little establishment already mentioned the eggs of trout have been brought from the Teste and Bourne in Hampshire, from the Colne in Herts, and the Wandle in Surrey. Salmon ova have been obtained from several parts of Ireland, as well as from the Rhine, the char have come from Geneva, and the grayling been taken from several British rivers where this delicate and beautiful fish survives. Eggs can be safely conveyed, if packed carefully in wet moss and placed in wooden boxes.

It will always be found advisable to make provision in various parts of the river which is intended to be the future residence of the young fish, not only for the youthful, but their adult state. Several fish, such as trout, pass solitary lives, each choosing some particular haunt, and only changing its residence when it has out-grown its home or can oust a weaker fish from some comfortable nook.

The trout loves to lie under the shelter of large stones, and if a good artificial place of refuge can be made, the best fish are sure to come and take possession of it. Perhaps the very best substance for this purpose is the semi-vitrified brick which is found in kilns after the burning, and which goes by the name of brick-burrs. This substance is in rather large masses, very irregular, and not only affords a home which no sensible trout will despise, but is an effectual barrier to the use of the net, serving the same purpose in the river as ‘ bushing ’ in the open fields.

Feeding the trout is also useful, for it teaches the fish to remain near the same spot, and has a marvellous effect towards increasing its growth. Scarcely any creature, and certainly no fish, repays care and good feeding better than the trout, two pounds having been added to the weight of a fish during a single summer. When the trout attains a moderate size it will eat all kinds of animal substance, though it has a predilection for the great dew-worms that are found at night on the grass or gravel walks.

These great, fat, and wary creatures can be caught plentifully by searching for them at night by the aid of a bull's-eye lantern ; only the step of the hunter must be very quiet, as they are apt to slip back into their holes if alarmed. Should they not come readily to the surface, they may usually be induced to do so by driving the prongs of a garden-fork into the ground and working it about so as to shake the earth around; and, if they still should be obdurate, they may be brought to light by pouring over the ground some water in which a very little ammonia has been dissolved.

Not only the river fishes, but those of the salt water can be reared from the earliest stages, and kept in ponds and fattened just like chickens, only with much less trouble and expense. Even the flat fishes are capable of being thus fattened, and become wonderfully thick in body and firm in flesh, so that their weight is really astonishing when it is compared with their length. The food which they require is of the cheapest kind, and the fattened fish can be sold for so high a price as to render the speculation extremely remunerative.

A few lines must be given to the machine in which fish can be conveyed for great distances without suffering damage or perishing for want of air. It consists, as may be seen from the accompanying illustration, of a square metal box closed above with a cover of perforated zinc. The box is not nearly filled with water, so that there is little fear of the contents being splashed out by the shaking incident to all travelling. The fish congregate at the bottom, and would soon exhaust the air contained in the water were it not renewed by artificial means.

In one corner of the box is placed a forcing-pump, neatly fitted up with appropriate valves, and communicating with a tube which passes down the corner and crosses the bottom. The lower portion of the tube is pierced with holes like those of a watering-cart, and the pump is so arranged that at each stroke atmospheric air is driven through the tube and bubbles upwards through the water, vivifying the exhausted liquid in its progress. Little trouble is expended on the process, as

TRANSPORTING BOX.

half a dozen strokes only are needed at a time, and the pump is so lightly constructed that a child can work it.

The reader will observe that there is not the least mystery or even difficulty about the process, and that anyone who can obtain a supply of water is able to hatch and rear young fish until they are old enough to put into a river. Should the fish be of the non-migratory kind they may be placed in a pond, where they will grow with great rapidity, and are always at hand when needed. In one pond which had been thus stocked, and was netted three years after the tiny inmates were admitted, no less than eight thousand pounds weight of fish were captured by a single sweep of the net. This pond was near Montmirail, in the department of the Marne, and is now unfortunately cleared of fish, the proprietor having determined on filling it up and using the ground for agricultural purposes.

It is rather remarkable that the Chinese, who seem from time immemorial to have known the rudiments of almost every science, and never to have advanced beyond them, are well acquainted with the principles of pisciculture, and have carried out the science to a greater extent than is usual with that thrifty and omnivorous nation, except when a supply of food is in question. They have even discovered that when the little fishes have absorbed the egg vesicles, and are beginning to need food which cannot be supplied in the natural manner by casual insects and aquatic animalcules, the best way to feed them is to beat up the yolk of an egg and pour it into the water; thus furnishing them with a kind of diet that requires no trouble to procure, being carried into their tiny mouths by the mere action of the water; and which is analogous to the nutriment contained in the vesicle from which they had previously drawn their support.

It is of course impossible, in the limited space which can be allotted to a single subject in these pages, to give more than a superficial sketch of the processes employed in pisciculture, and a brief notice of the benefits which are likely to accrue to a nation which rightly practises the art.

In this country, where so much is left to individual enterprise, and so little is intrusted to centralisation, it is scarcely to be expected that the Government will take up the question. Therefore, although the subject is really one of national importance, it must rest on its own money-producing merits, like any other kind of merchandise; and all that can be at present done by the press is to show the ease with which a fish-hatching apparatus can be established, the very little capital which is sunk in its erection and management, and the very large return which is made in proportion to the sums invested therein.

THE HOME OF A NATURALIST.[1]

THERE are naturalists and naturalists. Certain dreadfully scientific persons who call themselves by that
name seem to consider zoology and comparative
anatomy as convertible terms. When they see a
creature new to them they are seized with a burning
desire to cut it up, to analyse it, to get it under the
microscope, to publish a learned work about it, which
no one can read without an expensive Greek Lexicon,
and to 'put up' its remains in cells and bottles. They
delight in an abnormal hæmopophysis ; they pin their
faith on a pterygoid process, they stake their reputa-
tion on the number of tubercles in a second molar
tooth, and they quarrel with each other about a notch
on the basisphenoid bone.

Then there are the 'field naturalists,' who delight
in penetrating to the homes and haunts of the creatures
which they love, and spend whole days and nights in
watching their habits. Sometimes a field naturalist
remains at home, and immortalises an obscure village
by the simple process of using his eyes and telling his

---

[1] The proof sheets of this article, which was written in 1862, were
corrected by Mr. Waterton.

friends what he has seen. Another wanders far abroad in quest of new wonders, and, if he faithfully narrates the marvels he has witnessed, may calculate on being put down by newspaper critics as a skilful archer with the long-bow. Such a man was Le Vaillant, and such his reception by the critical world.

' Giraffe ? Humbug ! ' was the general criticism.

' Contrary to the laws of nature,' said the scientific.

' Would be liable to nine feet of sore throat,' wrote the witty.

And so the critics and the public enjoyed themselves amazingly at the traveller's expense, until the Pacha of Egypt sent two living giraffes to Europe, and turned the laugh in the other direction.

Such a man was Bruce, and such his reception. Peter Pindar showered most pungent epigrams upon his devoted head, and assailed him with most unsavoury comparisons. Perhaps there has been no statement of any traveller that raised such a storm of ridicule as Bruce's perfectly true account of eating beef cut from the living ox :

> Nor have I been where men (what loss, alas !)
> Kill half a cow and turn the rest to grass,

writes the poet, who wisely kept out of the redoubtable traveller's way, unwilling to share the fate of a contemporary caviller, who avowed that to devour raw beef was impossible, and was compelled, at the point of the sword, to eat his own words, together with a raw and freshly cut steak. *Solvitur edendo.*

Such a man was Charles Waterton, and such his reception; the ride on the cayman's back being treated by the press after the same fashion as Le Vaillant's giraffe and Bruce's ox. Time, however, is the true critic; giraffes are now as familiar as donkeys; eating flesh 'in the blood' is now known to be a custom existing in many nations from time immemorial; and the ride on the cayman has long been deprived of all marvel, except as a bold and dashing method of securing a powerful animal without damaging the skin.[1]

The discoveries of Mr. Waterton in Guiana are too well known to need even a passing reference; but, though better known to fame, are quite equalled in importance by the perpetual labours of half a century employed in observing the habits of living beings of our own land, and restoring to the dead skin the energetic contours of the living form.

There is perhaps no place in England where the greatest natural advantages have been so promptly seized and so largely improved as at Walton Hall, and it really seems almost impossible for such a combination of favourable conditions to be elsewhere achieved. There are many devoted naturalists who would exult in

---

[1] A clergyman of my acquaintance had in his employ, while he was resident in Guiana many years ago, one of the men who were with Mr. Waterton when he captured the alligator. The singular four-pointed hook with which the reptile was caught was at Walton Hall, covered with the marks of the alligator's teeth, and within a few feet hung the skin of the creature itself.

laying out their little domains in a manner calculated to attract the creatures which they love, but it can hardly be expected that another naturalist would possess the enormous natural advantages to be found at Walton Hall, and be blessed with health to manage his hobby for fifty years. As a general rule, the sapling which a young man plants is inherited as a tree by his grandson, and it .is very seldom found that the same eye which designed the original plan is permitted to see the results in their full perfection. Such, however, is the case in this present instance, and it may easily be imagined that where an extensive domain is laid out expressly for one purpose, which has been perseveringly carried out through half a century, and ever directed by the same mind which planned the design, a successful result is almost a matter of course.

The object which Mr. Waterton proposed to himself in 1813, the year after he had returned from the wilds of Guiana, whither he had gone in 1812 in quest of wourali poison, was to offer a hearty welcome to every bird and beast that chose to avail itself of his hospitality, and by affording them abundant food and a quiet retreat to induce them to frequent a spot where they would feel themselves secure from all enemies, save those which have been appointed to preserve the balance of nature. Food is always procurable, and the quiet retreat has been obtained by watching the habits of the various creatures, and providing them with such accommodations as they would seek in the wild state.

Mead, hill and dale, have been laid out to suit the idiosyncrasies of various species ; and trees of different kinds have been planted in clumps, rows, or in solitary state, to attract the birds that love such localities.

A large lake studded with islands and surrounded by simple meadow land, drooping willows, or thick woods, has been given up to the aquatic members of the feathered tribes, and rapid babbling brooks are at the service of those birds which need the running stream. An ancient ivy-covered gateway upon the borders of the lake has been altered for the benefit of the feathered race, and in a single season seven pairs of jackdaws, twenty-four pair of starlings, four pairs of ringdoves, a pair of owls, together with smaller birds, such as blackbills, redbreasts, redstarts, sparrows, and chaffinches, have built their nests in the same old tower, within a few feet of each other, and without attempting to quarrel.

In order to exclude human and quadrupedal enemies, a lofty wall has been built in the manner of a ring fence, surrounding about 260 acres of ground, having the lake in the centre and the house upon an island in the lake. A large telescope is mounted in a room which commands the whole lake and a considerable portion of the grounds, so that the most distant birds can be watched as perfectly as if they were close at hand. The wall was finished in 1826, and immediately upon its completion the herons came and built in the park. These beautiful birds absolutely swarm in the domain, and may be seen standing on one leg through the greater portion of the day, steady and impassible as if carved in wood. In order to suit the habits of these birds a channel has been cut on the side of a slight hill, which directs the waters of a little spring into the lake, and along this rivulet the herons love to stand.

To wander in the precincts of this domain seems a return to the primitive ages of the world, when man, beast, and bird had no dread of each other, and moved peacefully in the same happy grounds. The shyest birds are so well aware of their security that they care no more for spectators than the London sparrows for passengers, and will almost suffer themselves to be touched before they take the trouble to fly away for a few yards.

No sooner does the owner show himself than there is a general rush in his direction, and great is the flapping of wings and welcome of eager voices. Birds crowd round him on all sides to snatch the expected morsel from his hand; and I have seen him walk up to a bull that was sleepily reposing, coolly sit on his ribs, and feed the great beast with bread out of his pocket.

All the birds that inhabit this spot are perfectly free to come and go as they like, but the feeling of absolute safety is so great an attraction that no precautions are needed to keep them within the walls. Even the mallards—those shy and wary birds, that test all the sportsman's craft to approach—come in great flocks to the lake. They swim in large companies on its smooth waters, they edge the banks as far as the eye can reach, and behave altogether as if they were ordinary tame ducks. In the evening they take wing for the Lincolnshire fens, feed during the night, and return to the lake by day-dawn. The first point that struck me on my arrival at the house was the wild cry and loud wing-clatter of vigilant water-birds, invisible in the darkness, but quick enough of sight and ear to detect the presence of a stranger.

The whole place literally teems with life. Sweep the meadows, the trees, and the waters with the telescope at any season of the year, and each spot toward which the glass is directed is as busy as a disturbed ant-hill. On the lake may be seen Egyptian and Canadian geese, mallards, teal, wigeons, pochards, golden-eyes, tufted ducks, geese, and shovellers; and the only regret in the mind of the owner is that there is no inlet of sea-water. Still the marine birds often pay a visit to the lake, and the black-coated cormorant has made quite a long stay in its precincts, fishing boldly in front of the house, and gobbling eels with the astonishing voracity of its race.

The water-hens and coots run about under the very windows of the house, and sundry other birds would follow their example were it not for the jealousy of a fine pair of Egyptian geese, who choose to consider the whole island, together with the house, as their especial property, and drive away all other birds as soon as they dare to set a foot within the sacred precincts. The magpies and jackdaws, however, are too cunning for the geese, and as soon as a mess of potatoes is thrown down for the legitimate owners, a jackdaw is sure to come sweeping in one direction and a magpie in another, and to snap up the choicest morsels in spite of all the hoarse threats and angry gesticulations of the geese.

One of the most curious results of these investigations is the absolute certainty with which any bird can be attracted to a given locality by providing it with a suitable spot for its nest.

For example, in the hope of inducing the starlings to build in the grounds, twenty-four holes were bored in the old gateway tower. The result was that twenty-four pairs of starlings took possession of the holes, made their nests, and hatched their young therein. Encouraged by this success, and being desirous of giving the handsome and useful starling a home, the kind naturalist built two towers for the especial accommodation of these birds. Each tower is set on a pedestal of solid stone, so made that it cannot be climbed by cats or rats, the bane of all nestlings, and is absolutely filled with chambers.

There is a specially ingenious arrangement about these towers, which enables the bird to gain access to her nest through an aperture only just large enough to admit her body, and at the same time permits the observer to examine the nest and eggs at his leisure. The entrance to each chamber is closed by a cube of stone, having one of the corners squared away. When the stone is in its place the starling gets to her nest through the channel left by the missing corner ; but as the entire stone is moveable it can be pulled out at will, and thereby exposes the whole interior. The starlings are now so tame that they have no objection to being watched, and even after the stone is removed the bird sits calmly serene on her eggs, following the intruder with a fearless gaze.

Even the jackdaw builds in a hole within five feet of the ground, and close to the path which forms the back entrance to the house. The servants generally peep at the jackdaw's nest as they pass to and fro on their avocations, but the bird cares nothing for them, and treats them with supreme unconcern. Owls, again, were desired near the house, and a chamber was prepared for them in the gateway tower already mentioned. The apartment was hardly completed when a pair of barn owls took possession of it, and the spot has ever since been tenanted by these birds. Similarly, the brown owl was attracted by a large hole cut in a decaying tree, and by means of these semi-domestic guests, many disputed points in their habits have been cleared up, and their characters freed from the reproaches to which they had been subjected by all previous writers on natural history.

Herons, again, as has already been mentioned, took up their abode as soon as the park wall was completed, exhibiting thereby a marvellous instinct, which made the birds who build on the tops of the loftiest trees to feel that their homes were securely guarded by a wall not one fourth so high as their trees, and which they could overpass without the least difficulty The azure-backed and ruddy-breasted kingfisher finds a congenial home on the banks, though driven from the surrounding country by the cruel gun, and lays its pearl-like eggs in their bony nest, or flashes like a blue meteor along the shore in happy immunity from the dread tube that awaits it without the protecting wall. Feeling themselves perfectly secure, the birds act with the full freedom of their natures, and, unaffected by the presence of an observer, perform all the duties of life, play their pretty pranks, and exhibit their individuality as unconcernedly as if they were in a desert island where the foot of man had never trod.

The opportunities of gaining knowledge on such subjects are therefore unequalled, and great benefits have been conferred on the world by the information that has been obtained. Putting aside the interesting character of the pursuit, and the gratification which it affords to the observer, its results have been of practical utility. By a long series of useful observations the imputations under which many birds laboured have been entirely disproved, and in many cases a bird which was systematically persecuted and slain by the farmer has been shown to be a positive friend to its ignorant murderer. Such birds, for example, as the rook and crow have been proved to confer immense benefits on the agriculturist by devouring the subterranean larvæ, which stealthily consume the roots of the various crops, and are all the more formidable from the invisible nature of their assaults. The woodpecker, fiercely execrated as a destroyer of the trees, has been a right good friend to the landowner, feeding itself on the minute insects that burrow into the bark or into the decaying wood, and never chipping out its curious tunnel except in a spot where corruption has begun and is the sure precursor of death.

The kestrel, again, that was once thought to rival the kite in its raids upon the poultry-yard, is now known to do good service by day as does the owl by night, feeding either on the larger and more destructive insects, or on the little field-mice that swarm in all cultivated ground, and, if allowed to increase, make a woful diminution in the harvest. All the tribe of small birds, again, have been proved as benefactors to ten times the value of their depredations. In this domain is no restriction. Any bird is welcome to establish itself wherever it can find a suitable spot, may go wherever it chooses, and may eat whatever it likes. Field, orchard, and garden are equally at its service, and it may peck off buds from the trees, eat the cherries and currants, steal the peas, or revel in the corn just as its inclination may direct.

To distinguish friends from enemies is one of the first maxims in warfare, and is of primary importance in our daily struggles with the soil. All nature is in arms on one side or the other, and every being, whether animate or inanimate, is fighting against mankind, unless with Jasonic skill he compels them to turn their weapons against each other, and by mutual battles to consummate his peaceful victories.

In our own country, where individual energies are permitted to develop themselves without any restrictions of the ruling powers, no particular results have happened from the prevalent misapprehension respecting the friends or foes of husbandry ; but in a neighbouring land, famous for its logical precision of thought, its pitiless deduction of conclusions from premises, and the all-pervading influence of the supreme authorities, the little birds being known to eat seeds, strip the spring branches of their buds, and rob the autumnal trees of their produce, were condemned as destructive to the tenants, and, by the inexorable logic of facts, doomed to death. Rewards were offered throughout the land for the heads of small birds, just as in olden and foolish times the British churchwardens offered twopence per dozen for the heads of sparrows, and the juvenile population that was too young, and the adult population that was too idle to work, soon gathered a goodly number of heads and duly received their reward.

The results, however, were different, owing to two causes: the one being the universality of the measure in the one country, and its partial enforcement in the other ; and the second being that whereas the judicial authorities abroad, after paying for the birds' heads, took care to destroy them, the parochial authorities at home, after paying for the same, threw them out of the vestry window into the road, whence they were thriftily picked up by the expectant pensioners, and sold three or four times over. The consequence has been that on the Continent the insects have increased to such a fearful extent that societies have been lately formed for the express purpose of reintroducing the small birds that were extirpated at such an expenditure of time and money; and guarding against their slaughter by cruel little boys who take them out of their nests and murder the fledglings with the refined barbarity of juvenile civilisation, or by betasseled, green-clad, game-bag carrying gunners, who 'pot' them in the hedges and consider themselves sportsmen. Yet the question has been definitely settled more than twenty years ago, and in the 'Home of a Naturalist' sundry birds that have long laboured under causeless obloquy have not only been acquitted of all evil doings, but unexpectedly received into the number of our friends.

It will be at once seen that if any bird be attracted by food and a quiet retreat it may be expelled by an opposite mode of treatment, so that a knowledge of habit enables us to attract or expel those birds which we know by repeated observation to be our friends or foes. The same maxim applies to quadrupeds, and is often beyond all value.

For example, the farmer is almost invariably keen in hunting down and killing every weasel, stoat, or polecat in the neighbourhood, and his barn walls are generally defiled with numerous carcases nailed upon them as trophies of their slayer's vigilance. Yet every weasel is worth an annual sack of corn to the farmer, even after deducting the value of the few chickens and ducklings which it may destroy. Marauding cats are far more destructive than weasels, and if a farmer could succeed in clearing the neighbourhood of kestrels, rats, and the weasel tribe, his harvests would make but a poor show. There is no more determined enemy of the rat than the weasel and all its tribe. A thousand barn rats are calculated to devour two hundred pounds' worth of produce per annum; and taking into consideration the extraordinary powers of multiplication possessed by this insatiate devourer, who eats with equal voracity corn, cheese, bread, and meat of all kinds, whether raw or cooked, clambers into the pigeon-houses, murders the young, and destroys the unhatched eggs—nibbles its way into the hen-roosts by night, and kills the poultry as they quietly sleep on their perches—finds the ducks' nests and depopulates them—it is evident that any creature which gives its services in the destruction of this prolific and expensive animal is cheaply repaid at the cost of two or three chickens per annum. Some of the metropolitan hotel-keepers pay a tolerable annual wage to professional ratcatchers, and find themselves well remunerated for their outlay, even though the price which they pay is at least a hundred times as much as a weasel asks for his unceasing work.

Here, then, is another case proving the absolute

money value of practical zoology. The armed men rise from the furrows, fierce, hungry, and destructive, disputing its possession with the new-comer; but we fling among them the stone placed in our hand by science, they turn their arms against each other, and those which survive the contest become our willing slaves.

Still taking the rat as our text, see how a practical knowledge of its habits enables us to expel it from any place where it may have injudiciously taken up its abode. I say 'injudiciously,' because rats are useful enough in their right place, and by devouring all kinds of garbage save us from many pestilential diseases. Granting, however, that they have established themselves in some spot where their company is undesirable, how are we to expel them? Simply enough. Make their quarters unpleasant, and let them find nothing to eat. This was the method observed at Walton Hall, where the rats had triumphantly revelled for many a year, while the legitimate owner of the house was battling with snakes and fever in the distant forests of Guiana. Finding their haunts liable to continual raids, and their supplies of food cut off, they left the inhospitable house in disgust, and when fairly out of it were debarred from re-entrance by the judicious application of stone and iron. Fifteen years were occupied in learning the habits of the rat with sufficient accuracy to attain this successful result, but, considering the benefit conferred by this knowledge, the time was by no means wasted.

As a general fact, the result of the half century's

observation tends to prove that Nature will, in all ordinary cases, preserve her own balance; but that when man alters the conditions he must ever be watchful of his experiment, or run a risk of ignominious failure. In the present instance the wall affords no bar to the ingress or egress of the feathered race, who are thereby restored as nearly as possible to their original state of freedom, and are enabled to build their nests and forage for their young without the interruptions which would check them in any other part of England. The result is that although a vast number of species congregate within the domain —enough, according to the popular prejudices, to devastate the gardens, destroy the crops, and kill all the game—there are few places where crops, fruit, and flowers are so luxuriantly abundant, or where the game is more plentiful.

Not so, however, with the wingless creatures that are enclosed within its limits. Being unable to pass the wall, they are in fact prisoners; their conditions have been altered, and they are no longer able to preserve the rightful balance of Nature, so that man, who has interfered with the regular course of events and deprived the creatures of their liberty, is forced to accommodate himself to the altered circumstances, or take the consequences of his intrusion.

For example, of all laws to be observed in this little kingdom, the most stringent was that no fire-arms should be discharged within the walls—a needful and thoughtful regulation, as nothing alarms birds so

x

thoroughly as the report of a gun, or is so likely to deprive them of the secure retreat so necessary for their well-being. Now it so happened that a number of rabbits were enclosed within the wall on its completion, and, for a time, they did little damage. But rabbits are nearly as prolific as rats ; and, in spite of those that were killed by weasels, stoats, and polecats, their numbers increased in arithmetical progression, and they became scarcely less hurtful to the crops than the rats themselves, the turnips being almost destroyed by their busy teeth. At last the mandate was issued for their extirpation, and for the first time for many years guns were fired and dogs roamed at large within the sacred precincts. Curiously enough the result of the firing was rather contrary to expectation. Confident through long immunity, the birds troubled themselves very little about the guns. At first they were much disturbed at the unwonted sounds, but soon appeared to discover that they stood in no danger, and sat looking at the keepers and dogs with amusing composure. Even the herons only moved away to the tree tops, and the mallards contented themselves with leaving the banks as the dogs approached, and swimming towards the middle of the lake, where they paddled about in conscious security.

We have now to pass to a collateral branch of the same subject, and to call attention to the wonderful museum of stuffed animals which adorns the 'Home of a Naturalist,' and which is prepared on a principle equally remarkable for boldness of conception, originality of design, the Promethean fire which it enables a skilful operator to infuse into lifeless and flaccid remains, and the marvellous success with which it handles the two most troublesome departments of taxidermy, namely, the extremities and the naked skin. By this process the flowing undulations and softness of the bare skin are reproduced as they existed during the life of the subject; to the ears are given their original expressive roundness without the least symptom of the wrinkled outline which they usually assume in stuffed animals; the nose and muzzle retain their pouting fulness without requiring to be amputated and replaced by wax, gutta-percha, or other substances, and the paws or feet present their natural form and expression. Moreover, there is no wire or woodwork within the skin, and the weight of a prepared animal is so trifling that a man could carry away a cow or two on his back, take a tiger under each arm, sling half a dozen snakes round his neck, and walk off with his load.

We are all familiar with stuffed birds and beasts. Some of us may have had the misfortune to see some special pet carried off by death, and to have sent it to a ' naturalist,' to be stuffed. And those who have committed this affectionate error are sure to be dissatisfied with the operator, and to appreciate the infinite difference between the soft, graceful outlines, the expressive attitudes, and the sleek glossy coat of their former favourite, and the stiff, gaunt, distorted form of the stuffed skin, with its round staring eyes, its withered ears, lips, and nostrils, and its mummified feet, which bear no more resemblance to the extremities of the living creature than Yorick's skull to the living face. Even in the best specimens that have been stuffed in the ordinary manner, the feet, paws, and tail are at least sure to be failures after a few years have elapsed, while the ears and all parts of the body where the skin is devoid of covering become more and more shrivelled as time passes on. It is true that defective taxidermy is not generally detected, simply because those who look at a stuffed lion, eagle, snake, or crocodile are not sufficiently familiar with the living beings to appreciate the short-comings of the prepared skin ; but if they were as thoroughly acquainted with those creatures as with their own pet dog, cat, or bird, they would be equally capable of comprehending the effect which a badly prepared skin has on a naturalist, grating on his mind like false harmonies on the ear of a musician.

It is evident that the fault in defective taxidermy is twofold—firstly, ignorance on the part of the operator, and, secondly, the insufficiency of the method which he employs. Putting aside the former and more obvious cause, let us see how it is that even a dog or cat cannot be stuffed so that the prepared skin shall look exactly like the creature while instinct with life. The fact is that the present system, as generally employed, is radically false, and can but produce imperfect results—the skin being arranged while moist upon an inner basis of yielding substance, such as tow, moss, hay, &c., and suffered to dry almost at random, all manipulation being confined to the exterior. Now it is the property of moist skin to contract during the process of desiccation, and should its thickness be in the least unequal the contraction is unequal too, and so drags itself out of shape, and the hair and feathers out of their just 'set.' The only method by which an artist can ensure a successful result is to keep the skin under his control during the whole period occupied in drying, and to be able to reduce a wrinkle or produce a protuberance at will. He can thus restore the precise aspect of the being under his hands; he can give the external indications of swelling muscle or hidden joint, and impart to a mere hollow shell of skin the energy of a breathing creature. Such examples of true taxi- dermy are now to be found in Mr. Waterton's museum, and, unfortunately, in no other place.

There stands Chanticleer, proud and defiant, his crested head flung aside as if listening to a rival's challenge, his hackles bristling round his outstretched neck, and his armed legs firmly planted, as if awaiting the onset of his foe. There sits the pheasant, glorious in the full richness of nuptial plumage, its soft sleek outlines and undulating curves contrasting beautifully with the fiery action displayed by the champion of the poultry-yard. Here is a barn owl, fast asleep, not sitting on a branch, as is the custom of most sleeping birds, nor tucking its head under its feathers, but standing bolt upright, its legs stiff, as if two wooden skewers had been thrust up them, and its whole aspect irresistibly reminding the spectator of a dozing cat. Now here is an example which shows the value of understanding a bird's habits before undertaking to stuff its skin. Few persons knew that the owl slept in this odd position until Mr. Waterton found it out, and having discovered this peculiar trait of character, he has indelibly impressed it upon the preserved specimen.

In the museum are more than five hundred specimens preserved by one hand, not including a vast number of crabs, lobsters, insects, and various other smaller creatures; the great zoological value of the collection being that every specimen is represented in some natural and characteristic attitude in which it has been observed by the operator. Thus we have the toucans, sitting with their air of serious gravity, and the pert little toucanets, balancing themselves on the branches in the oddest manner, the bill and tail approaching each other beneath the bough. Numbers of parrots and parrakeets are displayed in all the attitudes which those mercurial birds assume, spreading their beautiful wings for flight, climbing up the boughs with their hooked beaks, ruffling their feathers, and scolding each other lustily, and, in fact, wanting nothing but movement to seem gifted with life.

There are the lovely humming-birds poised on steady wing, hovering about the flowers, or seated in their wee nests, and looking up with their pretty air of innocent audacity at the supposed intruder. Not a feather is missing or out of place, not a speck of black is to be seen on the burnished gorgets, which literally blaze with ruby, emerald, and topaz, when the sunbeams shine on them. The woodpeckers are hard at work on their trees, the quail trips daintily over the grass, and the warblers sit at rest on the branches, or flutter their plumage as if filled with ecstasy at their own melodious carols.

The great coulacounara snake lies coiled in dreadful folds, his eyes dully gleaming under their brows, and his head idly reposing on the pillow of his own body. Venomous serpents are seen lurking amid the foliage, one quietly sleeping, another drawing back the angry head in readiness for the stroke, the forked tongue quivering and the threatening fangs erect, while a third is triumphantly bearing off a fluttering victim in its jaws, the birds around fleeing in dismay.

Turning from the feathered to the furred races, the specimens are quite as characteristic. A huge ant-bear prowls along, his bushy tail curled over his back, and enveloping him in a torrent of hair, and his long snout held close to the ground, as if in search of his insect prey. A sloth is seen ascending a branch, clinging firmly with all its limbs, stretching out its neck, and wearing that peculiar pitiful, wistful look so characteristic of the creature. The weasel is seen, not stuffed, as is the custom in the dealers' shops, straight and long-backed like a furred lizard, but with arched back, recurved neck, and head drawn snake-like to the shoulders, just as the little creature appears when suddenly alarmed and ready to jump in any direction at a moment's notice.

Perhaps the apes are the most surprising examples of successful preparation. Everyone knows how utterly unsatisfactory is a stuffed monkey, with its face shrivelled out of all shape and expression, and looking as if punched out of an old shoe, its withered fingers like knotted sticks covered with scorched parchment, and the total want of 'set' in its fur—defects which increase with time, and quite ruin the real value and true object of the specimen.

But here is a young chimpanzee, sitting with a negligently easy air on a cocoa-nut, and contemplating the landscape with the air of profound wisdom and grotesque melancholy so often seen in the few tail-less apes that have been brought to this country in a living state. The mouth and lips possess all the soft roundness of life; the ears, nose, and forehead have regained their wonted contour; and even the bare palms are so perfect that the little wrinkles caused by their half-closed state, as they lie negligently on the lap, are reproduced with marvellous fidelity.

Get the creature between yourself and the window, and you will see that it is perfectly hollow, so hollow, indeed, that the hands and face are translucent as letter-paper—even to the very finger-nails. There is literally nothing in it but air, the skin being hard and elastic as if made of horn. A 'chuck' under the chin, and off tumbles its head, so as to allow a full view of the interior. Lift the creature, and the hand flies up with its lightness, as when one takes up an empty ewer thinking it to be full of water.

Almost anything may be done with the skin when once prepared after this fashion. Should it be stuffed abroad, it may be cut in twenty pieces for the convenience of package, and put together again without a mark to betray the junctions. See, it can be crumpled between the fingers and squeezed like a sponge, returning to its original shape by the strange elastic firmness which the skin has now attained. It can be picked up by a pinch of hair and swung about without damage. Its fur can be rumpled and pulled about until it sticks out in all directions, and then replaced with a few strokes of a brush. It may be kicked downstairs, or flung from the top of the Monument, without showing a sign of ill-usage. It may be squeezed flat as a pancake, and, when the pressure is removed, will resume its shape with the elasticity of a hollow india-rubber ball.

But better still, it is totally impervious to insect foes; it has no unpleasant smell such as is found in skins stuffed after the ordinary fashion; there is no horrid arsenical soap to endanger the sight, impair the appetite, and loosen the teeth of the operator; the creature stands boldly in the open air, with a simple glass cover to keep the dust off, and there is no need of camphor or turpentine, whose oppressive vapours pervade our museums, and give direful headaches to the visitors. You may take at random any of the five hundred specimens, say a bird, put it in a box, together with moth-eaten furs, feathers, and blankets, with mite-covered insects, and with a pint or two of the terrible dermestes—that scourge of museums—prolific, sharp-toothed, and voracious, capable of devouring a case-full of birds in a marvellously short time, and leaving no relics of the once beautiful inmates except some wires and a little tow clinging to them. Put the chest aside for twenty years, and when the accumulated dust has been brushed off, the bird will be found bright and un-injured as when it was first placed in the box.

In fact, the apparently frail and perishable skin has been rendered so impregnable to all ordinary foes that it can only be injured by main force, fire, or water, and even in the latter case could be soon re-modelled into its former shape. To all appearance, indeed, the light and delicate fur and down are likely to outlast the edifice of stone and iron in which they are sheltered, and to be a more enduring memorial of their preserver than monuments of brass or cenotaphs of marble. It will be seen, too, that by the plan of employing the mere skin the whole of the body is set free for the purposes of the anatomist: no slight advantage in the case of a rare or choice specimen.

Such are the results, but what of the means?

Simple in the extreme. The tools required hardly deserve the name, for all these wondrous effects have been produced with a penknife, a lump of wax, half-a-dozen needles, and three or four wooden skewers. The process is so cleanly that it can be conducted in a drawing-room, without soiling the most delicate furniture; and we have had the pleasure of seeing the inventor engaged in the manipulation of a pheasant, just as a lady employs her fingers on the elaborate entanglement of thread, called by courtesy her ' work.'

In simple fact the *modus agendi* is pure modelling, the skin being used as the material, and reduced by art to the plastic state of sculptor's clay, a temporary stuffing being only placed within it to keep the skin moderately distended during the progress of its drying. The obedience of the material to the touch of the hand is almost incredible; and in the collection may be seen several specimens that have purposely been distorted into all kinds of strange shapes, in order to show the value of the process in the hand of a master. Frogs, toads, and lizards are grotesquely transmuted into caricatures of the human form; extraneous joints, limbs, claws, and horns sprout from unexpected places.

Perhaps the most striking of these transformations is the well-known nondescript, wherein the natural countenance of the Howler monkey has been changed with such forcible and telling fidelity into the face of a quaint and eccentric but genial-hearted old man, that many of those who visit the museum leave it under the idea that they have been contemplating the prepared skin of a 'native,' and one gentleman, on seeing an engraving of the object, took it for a portrait of the operator, and thought that Mr. Waterton must be a very odd-looking person. Less in dimensions, but not less amusing, are the bizarre forms wittily ticketed as Cancer zodiacus and Diabolus cœruleus, two ludicrous combinations of heterogeneous parts, belonging to all kinds of creatures ; and the various odd compositions that meet the eye are made with a marvellous ingenuity that surpasses even the far-famed Japanese mermaids (of which, by the way, I have examined several), and bewilders the casual visitor to such an extent that he is led to doubt whether the very staircase may not be a deception. These objects are only manufactured for the purpose of showing the perfection to which the art of skin-modelling can be brought, and the plastic nature of the material placed in the taxidermist's hands.

It has been suggested that the time consumed in completing one of these specimens—namely, seven or eight weeks for a creature as large as a leopard—would debar professional taxidermists from employing the system. But each specimen only requires about half an hour's work daily, so that, after the first start, an industrious operator can turn out as many specimens as under the present system. Mr. Waterton, for example, always has several skins in hand, in different stages of progress, and by giving a few minutes' labour to each specimen several times daily, keeps up a constant influx of new objects into his museum. It is very interesting to watch the advance of the skin through the operation, and to see how it gradually grows from a wet and almost shapeless mass into a form apparently instinct with life and energy, like a lump of clay under the sculptor's hand; and how the skin, at first loose and flaccid, gradually acquires firmness and plasticity, until at length it obeys the slightest touch of the operator's hand, and permits each feather or hair to be arranged according to his will.

There are one or two other modes of taxidermy which deserve a passing notice. In one method, for example, the operator removes the skin, takes a cast in plaster-of-Paris of the ' ecorchée,' and stretches the skin over the cast, thus ensuring for the time an exact copy of the original. Yet even this plan, despite of its ingenuity, is but partially and temporarily successful; for all skin will persist in contracting as it dries, and the operator cannot possibly give the thousand little elevations and depressions of the softer parts, on which depends so much of the true expression.

Another most ingenious plan is that which has been employed by Professor Sokolov, of the Imperial University of Moscow. By this process, which consists of injecting certain preservative fluids into the system, the whole substance is rendered impervious to decay, and even the expression of the features so perfectly retained that the first impression of a spectator is that the form has been modelled in wax. Even the natural elasticity of the flesh is partially preserved, and if it be pinched it will give to the pressure and return to the original form. Moreover, the whole organisation remains so unchanged that it is still suitable for the scalpel of the anatomist, and even the delicate fibres of the muscles retain their organisation. Marvellous as is this preparation, it is still faulty in the extremities, to which the preserving fluid appears not always to find free access, on account of the small diameter of the capillaries. It is, however, a very great advance on all former systems of embalming, and as its essential processes are only the work of a few hours, it bids fair to be invaluable to comparative anatomists, who can thus get large and valuable specimens from distant lands without the vast outlay in spirits and great consumption of space that have hitherto been necessary.

Take it all in all, we have at present no process of taxidermy which presents so many excellencies and so few defects as that which is invented and practised by Mr. Waterton; and after a careful examination of almost every interesting specimen of taxidermy in the kingdom, we cannot but think that a judicious combination of the two systems (however opposite they may seem) of Mr. Waterton and Professor Sokolov would be of infinite value to science, inasmuch as the whole of the creature would be made available for the museum or the dissecting-room.

LIFE IN THE OCEAN WAVE.

It really must be a pleasant life in the clear salt sea, at least if we may judge from the splendid aquarium at the Crystal Palace, where we are, so to speak, let down some ten or twelve feet from the surface, and are brought face to face with its inhabitants. Imagine a spacious corridor, more than two hundred feet in length, having one side filled with large single-paned plate-glass windows, and that corridor sunk bodily into the sea so that through the windows we can watch the fishes, crabs, lobsters, shrimps, and other denizens of the shores, all busily engaged in their several vocations. Such is the general idea presented to the visitor when he enters the aquarium, and how that effect is produced we shall presently see. Besides their glass windows, through which we look into the sea, there are a number of comparatively shallow open tanks, so arranged that the visitor is able to look down upon the inhabitants and se them through the surface of the water.

Let us now take a walk round the corridors and inspect the ocean as seen through the glass. Each opening looks into a rocky cavern inhabited by various creatures, whose names are duly placarded in front of the window. One of these caverns is tenanted by sea cray-fish, lobsters, and certain most brilliant fish, covered with zebra-like stripes of green and pink, and called by the name of wrasse. These fishes possess a splendour which is almost tropical, and it is really hard to believe that they can be inhabitants of our comparatively dull coasts. The lobsters keep themselves quiet, as a rule, and wedge themselves into crannies, with their large claws hanging down like the paws of a dancing-dog. The cray-fish, however, are very locomotive, and it is astonishing how different they look in the water, or in a fishmonger's shop. Generally, during the day-time, they, like the lobsters, keep quiet in their retreats ; from which, however, their long, straight, stout antennæ project like the bayonets of sentries. And indeed this is one of the principal use of these organs. The cray-fish does not possess the powerful claws which form the lobster's weapons, but the antennæ serve a similar purpose, though in a different way. They are covered with small sharp projections, and are constructed much on the same principle as the terrible shark-tooth spears of Mangaia. With these spears the cray-fish fences, so to speak, and can drive away the generality of its foes. Mr. Lloyd, the designer and superintendent of the aquarium, informs us that he has seen various fishes thus driven away, and that even the conger-eel, with all its activity, is baffled by these singular weapons.

Towards evening the cray-fish come out from their caves and walk on the sand. It is astonishing how high they stand on their legs, and how large they look. One of the oddest things connected with them is that the wrasses have taken a fancy to swim under the cray-fish as they walk along, just like dogs under carts, sometimes slipping away for a moment, but always coming back again and resuming their places as if performing an act of duty. It is worth while to look carefully along the back of the cray-fish, on which are sundry white filaments waving steadily to and fro. These are in reality the ' cirrhi,' or bristle-legs of the ocean barnacle, numbers of which have fixed themselves on the shell of the cray-fish, and there sweep the waters with their net-like legs, so as to procure the tiny morsels of nutriment which float unseen, and are removed by the most efficient of Nature's scavengers.

Another grotto is chiefly inhabited by prawns, and presents a most singular and beautiful picture of life below the waves. Every projecting ledge of rock is crowded with the semi-transparent bodies of prawns, while the light flashes in bright sparklets from their waving antennæ, or is reflected from their mobile eyes, which glare like living opals. Some are sitting or crawling on the sand, while one or two hang suspended in the water, as the kestrel hangs in air, surrounding themselves with the tiny ripples caused by the ceaseless movement of the fan-legs, which drive the water through their gills.

Y

Calm and peaceful as is the grotto, it can at once be transformed into a scene of the wildest excitement. At a sign from Mr. Lloyd an attendant disappears, and presently a shadow is seen on the surface of the water, and a small white object—a morsel of whelk—falls from the shadow, and sinks slowly through the water. At first no notice is taken, but in a second or two a prawn leisurely crosses the track of the still sinking morsel. Instantly it scents its food, starts into sudden activity, and follows on the scent just as a hound follows a fox. Another and another succeed, and it is remarkable that while the prawns seem to let the food pass before their eyes without taking any particular notice of it, no sooner do they perceive the scent of the water through which it has sunk than they are after it at once. Now comes a whole shower of chopped whelk or mussel, and in a few seconds the tank is filled with prawns, all in busy excitement, some tucking up, or rather, down, their food, and others carrying off the white morsels to a quiet resting-place where they can eat in peace. Much the same habit distinguishes the whelks themselves and other carnivorous molluscs, which can be allured out of their retreats by trailing a morsel of food near them. They do not see the food, but they smell it, and forthwith set off in search of the expected banquet.

The reader will have probably remarked that we have only noticed a few of those marine animals who act as food. There are plenty of others, including the oyster, the mussel, and the cockle, the latter remarkable for its long coral-like feet, and its wonderful powers of jumping. There are soles, plaice, skate, and a variety of flat fish, all of which have their particular ways. The sole, for example, has a specially wide-awake air about it, as it lies on the sand with its head raised, and its large eyes set on two projections so as to give it a wide scope of vision. As it moves through the water it glides in a succession of undulating curves of the most graceful character, the glittering white undersurface occasionally showing itself, as the creature pursues its graceful course, every movement exemplifying a line of beauty. A dead sole lying flat on a fishmonger's slab and a living sole in motion seem almost to be different beings, so utterly unlike are they.

Plaice and dabs act much like the sole; but the ray lies flat and motionless on the sand, from which it can scarcely be distinguished. A touch, however, at once rouses it, and sends it darting through the water at wonderful speed. It soon slackens its pace, and comes slowly floating back to its old quarters, in a manner which irresistibly reminds the observer of Tom Hood's humorous comparison of 'swimming a kite.' Then there are plenty of cod, and very pretty creatures they are, with their large eyes, spotted bodies, and the broad waved silver stripe along their sides. To see them in perfection a few living shrimps should be thrown into the water, when the fish dart at them, making their glittering sides flash and sparkle in the light. It really seems impossible to eat cod fish after seeing the creatures alive.

As to the whiting, they look almost too beautiful to live, far too beautiful to eat. Their bodies are as if made of mother-of-pearl, over which an opaline lustre is perpetually flitting, and they are so translucent that if another fish passes behind one of them, its presence, though not its outline, can be seen through the whiting's body. There is ample store of mullet, which are particularly valuable inhabitants of an aquarium, because they clear away the vegetation which is always liable to collect on the glass, and to render it so dim that nothing can be seen behind it. Many of the fish have become so familiar that when food is placed on the end of a stick and offered to them they will come and eat it, acting, as a bystander said, 'just like Christians.' Indeed, according to Mr. Lloyd, they are a great deal better than the general run of Christians, since they never quarrel for quarrelling's sake. It is true that if one of them be sick its brethren will eat it, but that is only in accordance with the conservative laws of Nature. Or two gentlemen may quarrel over a lady, and one combatant be killed, in which case he is sure to be eaten ; but that is only the struggle for existence. Or a large hermit crab, which is in a little shell, may eject a little hermit crab who lives in a large shell, and enforces a change of residence ; but that is only carrying out the law of self-preservation. Spite and malice, however, have no place in the dispositions of these marine dwellers, and their natural tendency is to peace and quiet.

As to the odd fish in the aquarium, they are without number and we can only mention one or two.

The first of these is the butterfly gurnard, so called from the pectoral fins, which, when fully expanded, look exactly like the wings of some gorgeous butterfly. They are fan-shaped, and edged with the most vivid light blue. The disc of the fin is soft brown, powdered with light blue spots; while nearly in the centre is an oval patch of deep satiny blue, on which are scattered a number of pearl-white spots. The rest of the fish is plain brown, so that the contrast between the colour of the butterfly-like fins and the dull brown of the body is absolutely startling.

Another odd fish is the lump-sucker, so called on account of its lumpish form—which many of the spectators refuse to acknowledge to be that of a fish—and its peculiar sucking apparatus on the breast. With this sucker it clings firmly to rock, glass, or any similar substance, and has a quaint mode of always bending its tail on one side, so as to shorten its body nearly one half, and to take off the leverage which might affect the holding powers of its sucker. In this attitude it looks as little like a fish as may be, neither attitude, form, nor colour being in the least fish-like according to the ordinary ideas of fish. We must not pass over the long-bodied marine sticklebacks, notable for their habit of building nests under water. One of these fish, with more ingenuity than discretion, utilised the end of an old rope that was hanging in the water, and wove the ravelled fibres into its nest, not knowing that ropes which are hanging in the water are likely to be pulled up again.

Nor must we forget the smooth blenny, which can crawl up rocks and remain out of water for a considerable time, and can be so easily tamed that it can be fed while lying in the hand. Nor its near relative, the viviparous blenny, which, as its name imports, does not lay eggs after the manner of fish generally, but saves time by producing living young. Some little time ago there was a great fuss made about some exotic fish that had been shown to be viviparous, while all the while we had on our own shores a perfectly familiar viviparous fish, which no one seemed to think at all wonderful. Distance, as usual, had lent enchantment to the view.

Perforce we must part with the fish and inspect the molluscs.

At the head of them all come the cuttles, about which so much sensational writing has been inflicted on the public. There are several cuttles in the aquarium, the largest of which is the octopus, which the newspapers *will* call the devil fish, whereas it is, as a general rule, a very harmless creature, the real devil fish being a gigantic skate, one of which has been known to measure twenty-eight feet in width and twenty in length, and to weigh a ton. The octopus is nothing but a big sea-slug, with a higher development than the generality of molluscs. In fact, in the cuttles we find the first approach to a brain and a skull, though the former is insignificant, and the latter only gristle. As to the eight arms of the creature, with their array of suckers, they are easily to be accounted for. Most of us have seen a snail crawling on glass, and noticed the flat 'foot' on which it glides. Now supposing this foot to be cut into eight strips, and each strip to be greatly elongated, it is evident that the holding power of the foot would be gone, so in order to compensate for this defect, each of the strips is furnished with suckers, sometimes in a single, and sometimes in a double row. All the stories that have been told about its power of leaving the water and attacking men on shore are as absurdly false as the old fable of the sailing nautilus—which, by the way, we lately saw reproduced in two modern illustrated books.

The cuttles in the aquarium have a great dislike to showing themselves, and generally spend their time in the rock crannies, where they cannot be seen. The large octopus had, on the occasion of our visit, been (*pace* Mr. Lloyd) sulky for weeks, and never would come out of his den, showing nothing but the tips of one or two arms writhing about the entrance of his home. However, as if conscious that he was going into print, he gallantly came out and exhibited all his points, just as if he had been a trained cuttle going through all his performances. First he slowly emerged from his hole, and let himself sink gently on the sand. Next he began to crawl, using his arms by way of legs, and ending by standing on two of them, while with the others he explored the rock he meant to scale. Then he showed us his skill in climbing, stretching the arms to their fullest length, fixing the suckers, and thus drawing his body up. That done, he suddenly shot through the water backwards, propelled partly by water ejected from a tube called a siphon, and partly by the contraction of the webbed base of the foot.

Finally he sank again to the sand, and went through a number of movements, which we can literally call evolutions, as they seemed to be intended for the purpose of showing the mobility of the arms, and the extraordinary and complicated curves and knots into which they could be thrown. Whether induced by example, or by the spirit of competition, we cannot say, but two other cuttles came from their hiding-places, and we had presented to us the unwonted spectacle of three living cuttles all in view at the same time. This was the more noticeable because an attendant had in vain tried to dislodge them. No endeavour of his had the least effect, but almost as soon as he had desisted they came out of their own accord. Very human indeed. At last the octopus concluded his performance by retiring to his old home, a hole into which he gradually packed himself in a manner that irresistibly reminded us of Baron Bradwardine insinuating himself into his 'Patmos.' What more can we say? We could fill column after column with the ways and manners of these curious beings—how the sea-anemones are fed regularly with bits of mussel, how the purple whelk persists in congregating under the water inlets, and there depositing its strange foot-stalked eggs, each egg looking very much like a ninepin set on end—how many of the lower beings, including our Darwinian ancestors the Ascidians, seem to come of their own accord, and are far better and cleaner for zoological purposes than those which have been taken in the sea. But space fails us, and we can but briefly describe how this vast army of animated beings is kept in health.

The key to this problem is double, namely, cleanliness and circulation.

The cleaning of the aquarium is attended to with the most scrupulous fidelity. The water passes through tubes of vulcanite, so that no metallic oxide can poison it. The food which is given to the animals is absolutely fresh, and that which has to be cut is chopped on a large sloping slate, over which runs a stream of water. No dead animal is allowed to remain in the tanks, and every particle of uneaten food is removed.

Circulation is as carefully secured. There is a double set of engines and pumps, by means of which the water is kept perpetually flowing through the tanks. By the courtesy of Mr. Lloyd we were taken behind the scenes, and enabled to understand the enormous amount of thought and labour which alone can maintain such an establishment. As Mr. Lloyd well remarks, the engines are the heart of the aquarium, and the water is the blood. Under the feet of the visitor is an enormous tank, in which are eighty thousand gallons of sea water, exclusive of the twenty thousand gallons contained in the inhabited tanks. By means of the engine the water is kept perpetually circulating through the upper and lower tanks, while at the same time air is driven forcibly into it, and it is exposed both to the fresh air and to sunshine. Everything is carefully calculated, and even the wooden bridge on which the attendants walk has another use, serving also to keep off the light from the water, and so to discourage the excessive growth of the green algæ which are the plague of most aquariums. The consequence of this constant care is that the water is kept bright and clear as crystal, and, though the visitor is surrounded with many thousands of living beings, not the least evil odour is perceptible, the air being as fresh and pure as it is outside the building.

Uniformity of temperature is also secured, and even during the few broiling days of last year, when the Greenwich register showed 88 degrees in the shade, the temperature of the aquarium was just 20 degrees lower. After all that we have said, can the reader do better than make a long visit to the aquarium, buy a handbook, and make himself thoroughly acquainted with the world of wonders that lies unknown at our feet as we wander on the seashore ?

OUR LAST HIPPOPOTAMUS.

THE poor little hippopotamus is dead.   It was scarcely expected to live; but its death was nevertheless a severe disappointment, especially after the trouble and personal risk that were involved in the attempt to save its life.

About five o'clock in the morning the little animal was born.   The keeper knew that it was there from the odd sounds made by its mother— sounds of angry jealousy against some foe unknown. She slapped her vast jaws together, gnashed her teeth, and snorted loud defiance; though no one was in the house except the keeper, who was watching her from his unseen post of observation above. As daylight broke the small hippopotamus was seen lying by its mother; and the two were anxiously watched in order to find out whether the young one took nourishment. This it was never seen to do. It followed its mother about wherever she went, so that it was not deficient in strength; but it was never once seen to suck. Still, though in the daytime it certainly took no nourishment, it is impossible to say whether or not it may have done so at night; for many of those animals are exceedingly shy and wary, and will not even feed if they think that they can be seen.

In the present case, however, there is scarcely any doubt that the little hippopotamus took no nourishment at all. Either it did not know where to seek food, or its mother was too stupid to show it, for it did attempt to extract milk from her ears, but naturally failed. Two days having thus passed, it was evident that unless the young one could be separated from its mother and artificially fed it must inevitably die; but the difficulty lay in the mode of separation. To rob a mother of her young is proverbially dangerous; and when the mother is a savage, cross-grained beast, weighing some three tons, and capable of 'chawing up a human' with the greatest ease, the task is peculiarly perilous.

On Tuesday, the 9th, an attempt was made to get the mother away from her child. The two were lying on the ground, at some little distance from the water; and it was thought that if the mother could be decoyed into the pond, it might be possible to abstract the young calf before she could get out again. Now this creature is emphatically a good hater. She hates all kinds of things and persons. She hates workmen without their coats. I ónce saw her charge at a workman, and bite at the iron bars so savagely that she broke one of her enormous teeth completely into the jaw. But if there be one thing she hates beyond all it is the garden engine. The very sight of it, or even the sound of its wheels, sets her beside herself with rage, and whenever she sees it she is sure to charge. Accordingly the engine was run towards the water. In went the hippopotamus ; but unfortunately the little one plumped in after its mother, so that the ruse failed of its effect. The calf, in spite of its tender age, evidently enjoyed the water very much, swam about, and finally went to sleep, with its chin resting on the side of the pond—a favourite attitude with these animals.

Next day, when the mother walked about the house, her child lay still, being evidently weaker ; so that Mr. Bartlett, the superintendent of the Gardens, decided on making another attempt to get the calf out of the house. He got together a small but effective staff, and laid out his plans. In order that the public might realise the difficulty of the situation, the perilous nature of the task, and the ingenuity of the device, I must briefly describe the scene of operations. On the right of the den is a small pond, shut off from the platform at will by iron railings ; and on the extreme left is a small door, barely thirty inches square. The scheme was as follows. The garden engine was to be again run into the building, and so soon as the hippopotamus charged into the water, one of the men was to dash into the cage—if possible, to shut the animal into the pond, get out the young one, and make his escape before the mother could reach him.

Accordingly Mr. Bartlett stood by the door, ready

to open it at the right moment; his son, with
Thompson and Godfrey, two of the keepers, was close at
hand with a stout cloth; Prescott, the head keeper,
was in charge of the engine, and Scott, the elephant-
keeper—a very active and daring man—was selected
for the dangerous task of entering the den. Everything was arranged, to use Mr. Bartlett's own expression, 'like a trick in a pantomime;' for the whole business could not last more than a few seconds.

All being ready, Prescott ran the engine into the
house and began to pump into the pond, that being the
insult which the hippopotamus will least of all endure.
In she went, and, as she rose, Prescott pumped the
water in her face; thus half blinding her, and gaining
just the time that was needed. Simultaneously Scott
ran into the den, picked up the young calf, which was
lying close to the water, and handed it through the
door to Mr. Bartlett and his assistants. The expectant
party took it in the cloth, and ran away with it. Mr.
Bartlett fastened the door, Prescott ran the engine out
of the house; and, before she recovered from the surprise of the water in her face, the enraged hippopota-
mus· found herself alone, with nothing on which to
vent her fury. Of course she missed her calf, and
began to hunt for it, but her rage soon cooled down when
she could see no enemy; and, except bearing a rather
deeper grudge than usual against Mr. Bartlett, she regained her usual temper. Still when I paid her a visit
she looked ominously sulky; and, as she lay half in and

half out of the water, her eye had a wicked glare in it that was not pleasant to see.

Without a practical knowledge of the animal it is scarcely possible to realise the difficulty and danger of the task which was so successfully performed. Under any circumstances it is not easy to pick up a hippopotamus, however young. It weighs somewhere about a hundred pounds, and its skin is as slippery as if it had been dipped in oil. Add to this, that the floor of the den is wet and smooth, offering scarcely any foothold; that the young captive kicks and yells with all its power; and that within a few feet is its infuriated mother—and some idea may be formed of the feat which was achieved by those six men. Then no one who has not seen the hippopotamus in one of her furies can appreciate the risk that is run by any man who goes into the den. She is quick and active beyond conception, flies about like lightning, bellowing forth roar after roar and making the building tremble. On the present occasion her deafening roars did good service; for they completely drowned the cries of her young one, and enabled the keepers to carry it off unheard as well as unseen. She can rear herself on her hind legs in her attempts to get at a supposed enemy; and when her weight of three tons is brought to bear upon railings which are not too strong, no small nerve is required in treating such an animal.

Now let us follow the fortunes of the calf. It was taken quite to the other end of the Gardens, that its

mother should not be able to hear its cries. Had she
done so, she would have tried to force her way to its
rescue, and it is very doubtful whether, in that case, the
iron bars of the den, or even the wall of the building
itself, could have withstood the shock of her reckless
charges. When comfortably housed, the little animal
was found to be suffering severely from cold, and
means were at once devised to restore the proper temperature. Blankets were dipped in boiling water,
wrung as dry as possible, and then rolled round the
sufferer. Over the blankets was laid a thick coat of
cotton wool, and before very long Mr. Bartlett had the
pleasure of finding the heat of the body fairly restored.
Nourishment was the next business. At first every
offer was refused, but by degrees the calf was induced to
suck at an ingenious apparatus of india-rubber and
canvas attached to the mouth of a bottle filled with warm
milk. It was found necessary to blind its eyes when
the bottle was placed to its lips, and to preserve absolute silence, for the little creature was so sensitive that
it would take no nourishment so long as it could see a
human being or hear the sound of a human voice. Its
extreme shyness gave some colour to the belief that it
may have sucked in the night, when no one could see
it. The warm milk seemed to comfort the animal, and
it soon quieted down and slept.

The milk, by the way, was chiefly that of the ass,
for, in an early stage of life, cow's milk is found to be
too rich, and to be therefore fatal. This was shown by

the experience gained at Amsterdam. Five hippopotamus calves have been born there, and nearly all have died soon after birth. On dissection after death it was found that the first two calves had died from indigestion, the cow's milk with which they were fed having been curdled into solid lumps. Afterwards ass's milk was used, and with sufficient success to enable a young hippopotamus to be reared. There seems to be an adverse fate against the hippopotamus. This European specimen was bought, when a few months old, by an American firm. They gave 1,000*l* for it, and thought partially to 'recoup' themselves by showing it in England before it crossed the Atlantic. The speculation was unfortunate, for the animal was burned to death in the disastrous fire which destroyed the Tropical Department of the Crystal Palace some years ago, so that there is not a single specimen of a hippopotamus bred in Europe.[1]

To return to our calf. It took about three pints of milk in six hours, once imbibing nearly a pint at a time. Nothing, however, could save it, and about seven p.m. of the same day it died. Should the reader wish to know what a hippopotamus baby looks like, he has only to go to the Zoological Gardens, where he may see the stuffed skin of the older calf, and close to it a plaster cast of the baby, taken by Mr. Frank Buckland.

There is a comic element in most human affairs;

---

[1] The reader is probably aware that since this article was written, the rearing of the young Hippopotamus has been successfully accomplished.

and this was furnished by the letters with which Mr. Bartlett was inundated, all giving advice as to the best mode of feeding the young calf. One correspondent proposed that milk should be squirted at its mouth from a syringe. Several suggested that the mother should be chloroformed, and the calf removed while she was insensible. Now anyone with the least knowledge of. chloroform is aware that to place an animal under its influence is a matter of the greatest difficulty. For example, some little time ago, when it was needful to put a tapir under chloroform, the operation lasted for an hour, and required the combined efforts of Mr. Bartlett and six assistants, acting under the direction of the surgeon, though they had an apparatus which fitted on the animal's head. As to chloroforming an hippopotamus, it would be quite as easy to chloroform a whale. Yet the correspondents—all anonymous— gave their advice as if nothing could be easier. ' Chloroform the dam and take away the cub.' Perhaps the drollest of all the suggestions was a scheme for burning brimstone in the house until the mother should be stupified, and then removing the calf. How the calf was to be rendered sulphur-proof, or how the keepers were to breathe in an atmosphere which stupi- fies an hippopotamus, were points which the writer did not elucidate.

It is impossible to go behind the scenes, so to speak, of such an establishment as the Zoological Gardens, and to not admire the profound knowledge of beast nature—

the careful forethought, dauntless courage, and promptitude of action which are shown by those who have to keep within due bounds such ferocious and powerful brutes. It is the triumph of man's intellect over the instinct of the beast. Here we have one of the largest, fiercest, worst-tempered, most powerful animals on the face of the earth utterly conquered by a few men, all of whom it could tear to pieces if it only knew how strong it is and how weak they are. They employ no weapons —they use no terrorism; and yet this dread beast is absolutely powerless in their unarmed hands. So it is throughout the whole system.

For example, just before the birth of the young hippopotamus there was a very difficult business with the polar bears. A young male—quite a child—had been lately admitted. When he was allowed to join the original inhabitant he behaved himself very badly—snarling, growling, and altogether making himself an abominable nuisance to an elderly and quiet bear, who only wanted to be let alone. Presently he got into the water and swam about merrily: but after a little time he was evidently in difficulties. He could not get out again. His hind legs were too short to help him, and his forepaws could not hold the smooth stone. Moreover, his fur coat was dragging him down. When a polar bear's coat is clean, it throws off water like a duck's back, and the beast has a perfectly dry skin, though he may have been in the water for a long time; but if it be dirty it sucks in water like a sponge, so that its weight is increased three or four fold, and is extremely embarrassing when the animal is trying to crawl from the water up a smooth surface. Seeing that the bear was losing both strength and confidence, Mr. Bartlett let the water out of the pond, and then, by means of two poles which gently kept pressing its neck and knees, coaxed him backwards into the cage, in which he was shut.

Next day he was again admitted to the large den, when an amusing scene took place. So soon as he entered the old bear opened her mouth as widely as she could, and walked up to him, without uttering a sound. The action was expressive, and meant this : ' Look here, young gentleman—these are my teeth, and if you behave to-day as you did yesterday you will get a taste of their quality.' He took the advice, and demeaned himself in a sober and respectful manner. Most animals act in the same way. If, for example, a new monkey is put into the great cage, there is a general showing of teeth, accompanied by mutual comparison of strength and consequent amity :

> Marry, peace it bodes, and love, and quiet life,
> And awful rule, and right supremacy.

Here, as in the aquarium, there is no quarrelling for mere malice, and, indeed, there is but little squabbling on any account. The weaker acknowledge the power of the stronger, and there is an end of the matter at once.

Another point that has to be kept in mind is the extreme sensitiveness of the animals; and it is a remarkable fact that those creatures which have been born in the Gardens are much more difficult to manage than those which have been captured and imported. The truth is that the latter have been inured to many changes by reason of their travels; whereas the former have never seen anything but their own den or yard, and are horribly frightened at the merest trifle to which they are not accustomed. We believe that if one of the lions which have been born in the Gardens were to escape among the Sunday visitors, not a soul of them would be half so much frightened as the lion. It is really astonishing what a consternation is sometimes aroused by the most trivial cause. Not long ago Mr. Bartlett went as usual into the giraffe house. The animals at once flew into a state of the most violent excitement, dashed about the house, and seemed likely to break their necks. Mr. Bartlett at once divined the cause of their terror; he was wearing slippers instead of boots, and his noiseless movements struck them with a sense of mysterious dread. They had always been accustomed to hear as well as see a human footstep, and the absence of noise filled them with ungovernable terror. So he stamped as loud as he could, spoke to them, and they immediately calmed down.

Thus it is that the management of the Gardens is conducted. The disposition of every animal is carefully studied, and the keeper tries, so far as he can, to place himself in the mental position of his charge, and to anticipate its thoughts. Violence is never used. It would not only be futile at the time, but it would destroy all hope of obtaining future obedience. As it is, the animals find that in some mysterious manner they are continually obeying the will of their keeper, and they get by degrees into a habit of obedience more or less perfect, according to the nature of the particular species and the disposition of the individual. Knowledge is power here. It is equally exercised over the largest and the smallest animals in the place; and whether the keeper be in charge of a harvest-mouse which will scarcely balance a halfpenny in the scales, of an elephant nine feet high, or an hippopotamus weighing three tons, the human intellect equally asserts itself, and all equally acknowledge its sway.

LONDON: PRINTED BY
SPOTTISWOODE AND CO., NEW-STREET SQUARE
AND PARLIAMENT STREET

MARCH 1877.

GENERAL LIST OF WORKS

PUBLISHED BY

Messrs. LONGMANS, GREEN, and CO

PATERNOSTER ROW, LONDON.

———o•o°•o°o•o———

History, Politics, Historical Memoirs, &c.

SKETCHES of OTTOMAN HISTORY. By the Very Rev. R. W. CHURCH, Dean of St. Paul's. 1 vol. crown 8vo. *[Nearly ready.*

The **EASTERN QUESTION.** By the Rev. MALCOLM MacCOLL, M.A. 8vo. *[Nearly ready.*

The **HISTORY of ENGLAND** from the Fall of Wolsey to the Defeat of the Spanish Armada. By JAMES ANTHONY FROUDE, M.A. late Fellow of Exeter College, Oxford.

> LIBRARY EDITION, Twelve Volumes, 8vo. price £8. 18*s.*
> CABINET EDITION, Twelve Volumes, crown 8vo. price 72*s.*

The **ENGLISH in IRELAND** in the **EIGHTEENTH CENTURY.** By JAMES ANTHONY FROUDE, M.A. late Fellow of Exeter College, Oxford. 3 vols. 8vo. price 48*s.*

The **HISTORY of ENGLAND** from the Accession of James the Second. By Lord MACAULAY.

> STUDENT'S EDITION, 2 vols. crown 8vo. 12*s.*
> PEOPLE'S EDITION, 4 vols. crown 8vo. 16*s.*
> CABINET EDITION, 8 vols. post 8vo. 48*s.*
> LIBRARY EDITION, 5 vols. 8vo. £4.

LORD MACAULAY'S WORKS. Complete and Uniform Library Edition. Edited by his Sister, Lady TREVELYAN. 8 vols. 8vo. with Portrait, price £5. 5*s.* cloth, or £8. 8*s.* bound in tree-calf by Rivière.

On **PARLIAMENTARY GOVERNMENT in ENGLAND;** its Origin, Development, and Practical Operation. By ALPHEUS TODD, Librarian of the Legislative Assembly of Canada. 2 vols. 8vo. price £1. 17*s.*

The **CONSTITUTIONAL HISTORY of ENGLAND,** since the Accession of George III. 1760–1860. By Sir THOMAS ERSKINE MAY, K.C.B. D.C.L. The Fifth Edition, thoroughly revised. 3 vols. crown 8vo. price 18*s.*

DEMOCRACY in EUROPE; a History. By Sir THOMAS ERSKINE MAY, K.C.B. D.C.L. 2 vols. 8vo. *[In the press.*

JOURNAL of the REIGNS of KING GEORGE IV. and KING WILLIAM IV. By the late CHARLES C. F. GREVILLE, Esq. Edited by HENRY REEVE, Esq. Fifth Edition. 3 vols. 8vo. 36*s.*

▲

The OXFORD REFORMERS — John Colet, Erasmus, and Thomas More : being a History of their Fellow-work. By FREDERIC SEEBOHM. Second Edition, enlarged. 8vo. 14*s*.

LECTURES on the HISTORY of ENGLAND, from the Earliest Times to the Death of King Edward II. By WILLIAM LONGMAN, F.S.A. With Maps and Illustrations. 8vo. 15*s*.

The HISTORY of the LIFE and TIMES of EDWARD the THIRD. By WILLIAM LONGMAN, F.S.A. With 9 Maps, 8 Plates, and 16 Woodcuts. 2 vols. 8vo. 28*s*.

INTRODUCTORY LECTURES on MODERN HISTORY. By THOMAS ARNOLD, D.D. 8vo. price 7*s*. 6*d*.

WATERLOO LECTURES ; a Study of the Campaign of 1815. By Colonel CHARLES C. CHESNEY, R.E. Third Edition. 8vo. with Map, 10*s*. 6*d*.

The LIFE of SIMON DE MONTFORT, EARL of LEICESTER, with special reference to the Parliamentary History of his time. By GEORGE WALTER PROTHERO, M.A. With 2 Maps. Crown 8vo. 9*s*.

HISTORY of ENGLAND under the DUKE of BUCKINGHAM and CHARLES the FIRST, 1624–1628. By SAMUEL RAWSON GARDINER, late Student of Ch. Ch. 2 vols. 8vo. with Two Maps, price 24*s*.

The PERSONAL GOVERNMENT of CHARLES I. from the Death of Buckingham to the Declaration of the Judges in favour of Ship Money, 1628–1637. By S. R. GARDINER, late Student of Ch. Ch. 2 vols. 8vo. [*In the press.*

The SIXTH ORIENTAL MONARCHY; or, the Geography, History, and Antiquities of PARTHIA. By GEORGE RAWLINSON, M.A. Professor of Ancient History in the University of Oxford. Maps and Illustrations. 8vo. 16*s*.

The SEVENTH GREAT ORIENTAL MONARCHY; or, a History of the SASSANIANS: with Notices, Geographical and Antiquarian. By G. RAWLINSON, M.A. Map and numerous Illustrations. 8vo. price 28*s*.

ISLAM under the ARABS. By ROBERT DURIE OSBORN, Major in the Bengal Staff Corps. 8vo. 12*s*.

A HISTORY of GREECE. By the Rev. GEORGE W. COX, M.A. late Scholar of Trinity College, Oxford. VOLS. I. & II. (to the Close of the Peloponnesian War). 8vo. with Maps and Plans, 36*s*.

GENERAL HISTORY of GREECE to the Death of Alexander the Great; with a Sketch of the Subsequent History to the Present Time. By GEORGE W. COX, M.A. With 11 Maps. Crown 8vo. 7*s*. 6*d*.

The HISTORY of ROME. By WILLIAM IHNE. VOLS. I. and II. 8vo. price 30*s*. The Third Volume is in the press.

GENERAL HISTORY OF ROME from the Foundation of the City to the Fall of Augustulus, B.C. 753—A.D. 476. By the Very Rev. C. MERIVALE, D.D. Dean of Ely. With Five Maps. Crown 8vo. 7*s*. 6*d*.

HISTORY of the ROMANS under the EMPIRE. By the Very Rev. C. MERIVALE, D.D. Dean of Ely. 8 vols. post 8vo. 48*s*.

The FALL of the ROMAN REPUBLIC ; a Short History of the Last Century of the Commonwealth. By the same Author. 12mo. 7*s*. 6*d*.

The STUDENT'S MANUAL of the HISTORY of INDIA, from the Earliest Period to the Present. By Colonel MEADOWS TAYLOR, M.R.A.S. M.R.I.A.. Second Thousand. Crown 8vo. with Maps, 7*s*. 6*d*.

INDIAN POLITY; a View of the System of Administration in India. By Lieutenant-Colonel GEORGE CHESNEY, Fellow of the University of Calcutta. 8vo. with Map, 21*s.*

The **HISTORY of PRUSSIA,** from the Earliest Times to the Present Day; tracing the Origin and Development of her Military Organisation. By Captain W. J. WYATT. Vols. I. and II. A.D. 700 to A.D. 1525. 8vo. 36*s.*

The **CHILDHOOD of the ENGLISH NATION**; or, the Beginnings of English History. By ELLA S. ARMITAGE. Fcp. 8vo. 2*s.* 6*d.*

POPULAR HISTORY of FRANCE, from the Earliest Times to the Death of Louis XIV. By ELIZABETH M. SEWELL, Author of 'Amy Herbert' &c. With 8 Coloured Maps. Crown 8vo. 7*s.* 6*d.*

LORD MACAULAY'S CRITICAL and **HISTORICAL ESSAYS.** CHEAP EDITION, authorised and complete. Crown 8vo. 3*s.* 6*d.*

CABINET EDITION, 4 vols. post 8vo. 24*s.*	LIBRARY EDITION, 3 vols. 8vo. 36*s.*
PEOPLE'S EDITION, 2 vols. crown 8vo. 8*s.*	STUDENT'S EDITION, 1 vol. cr. 8vo. 6*s.*

HISTORY of EUROPEAN MORALS, from Augustus to Charlemagne. By W. E. H. LECKY, M.A. Third Edition. 2 vols. crown 8vo. price 16*s.*

HISTORY of the RISE and INFLUENCE of the SPIRIT of RATIONALISM in EUROPE. By W. E. H. LECKY, M.A. Fourth Edition. 2 vols. crown 8vo. price 16*s.*

The **NATIVE RACES of the PACIFIC STATES of NORTH AMERICA.** By HUBERT HOWE BANCROFT. 5 vols. 8vo. with Maps, £6. 5*s.*

HISTORY of the MONGOLS from the Ninth to the Nineteenth Century. By HENRY H. HOWORTH, F.S.A. VOL. I. *the Mongols Proper and the Kalmuks;* with Two Coloured Maps. Royal 8vo. 28*s.*

The **HISTORY of PHILOSOPHY,** from Thales to Comte. By GEORGE HENRY LEWES. Fourth Edition. 2 vols. 8vo. 32*s.*

The **MYTHOLOGY of the ARYAN NATIONS.** By GEORGE W. Cox, M.A. late Scholar of Trinity College, Oxford, 2 vols. 8vo. 28*s.*

TALES of ANCIENT GREECE. By GEORGE W. Cox, M.A. late Scholar of Trinity College, Oxford. Crown 8vo. price 6*s.* 6*d.*

HISTORY of CIVILISATION in England and France, Spain and Scotland. By HENRY THOMAS BUCKLE. Latest Edition of the entire Work, with a complete INDEX. 3 vols. crown 8vo. 24*s.*

SKETCH of the HISTORY of the CHURCH of ENGLAND to the Revolution of 1688. By the Right Rev. T. V. SHORT, D.D. sometime Bishop of St. Asaph. Eighth Edition. Crown 8vo. 7*s.* 6*d.*

EPOCHS of ANCIENT HISTORY. Edited by the Rev. G. W. Cox, M.A. and jointly by C. SANKEY, M.A. Ten Volumes, each complete in itself, in fcp. 8vo. with Maps and Indices :—

> BEESLY'S Gracchi, Marius, and Sulla, 2*s.* 6*d.*
> CAPES'S Age of the Antonines, 2*s.* 6*d.*
> CAPES'S Early Roman Empire, 2*s.* 6*d.*
> Cox's Athenian Empire, 2*s.* 6*d.*
> Cox's Greeks and Persians, 2*s.* 6*d.*
> CURTEIS'S Rise of the Macedonian Empire, 2*s.* 6*d.*
> IHNE'S Rome to its Capture by the Gauls, 2*s.* 6*d.*
> MERIVALE'S Roman Triumvirates, 2*s.* 6*d.*
> SANKEY'S Spartan and Theban Supremacy.   } *In the press.*
> SMITH'S Rome and Carthage, the Punic Wars.   }

EPOCHS of MODERN HISTORY. Edited by E. E. MORRIS, M.A. J. S. PHILLPOTTS, B.C.L. and C. COLBECK, M.A. Eleven volumes now published, each complete in itself, in fcp. 8vo. with Maps and Index :—

> Cox's Crusades, 2*s*. 6*d*.
> CREIGHTON'S Age of Elizabeth, 2*s*. 6*d*.
> GAIRDNER'S Houses of Lancaster and York, 2*s*. 6*d*.
> GARDINER'S Puritan Revolution, 2*s*. 6*d*.
> GARDINER'S Thirty Years' War, 2*s*. 6*d*.
> HALE'S Fall of the Stuarts, 2*s*. 6*d*.
> LUDLOW'S War of American Independence, 2*s*. 6*d*.
> MORRIS'S Age of Queen Anne, 2*s*. 6*d*.
> SEEBOHM'S Protestant Revolution, 2*s*. 6*d*.
> STUBBS'S Early Plantagenets, 2*s*. 6*d*.
> WARBURTON'S Edward III. 2*s*. 6*d*.

*** Other Epochs in preparation, in continuation of the Series.

REALITIES of IRISH LIFE. By W. STEUART TRENCH, late Land Agent in Ireland to the Marquess of Lansdowne, the Marquess of Bath, and Lord Digby. Cheaper Edition. Crown 8vo. price 2*s*. 6*d*.

MAUNDER'S HISTORICAL TREASURY; General Introductory Outlines of Universal History, and a series of Separate Histories. Latest Edition, revised by the Rev. G. W. Cox, M.A. Fcp. 8vo. 6*s*. cloth, or 10*s*. 6*d*. calf.

CATES' and WOODWARD'S ENCYCLOPÆDIA of CHRONOLOGY, HISTORICAL and BIOGRAPHICAL. 8vo. price 42*s*.

Biographical Works.

The LIFE and LETTERS of LORD MACAULAY. By his Nephew, G. OTTO TREVELYAN, M.P. Second Edition, with Additions and Corrections. 2 vols. 8vo. with Portrait, price 36*s*.

The LIFE of SIR WILLIAM FAIRBAIRN, Bart. F.R.S. Corresponding Member of the National Institute of France, &c. Partly written by himself; edited and completed by WILLIAM POLE, F.R.S. 8vo. Portrait, 18*s*.

ARTHUR SCHOPENHAUER, his LIFE and his PHILOSOPHY. By HELEN ZIMMERN. Post 8vo. with Portrait, 7*s*. 6*d*.

The LIFE and LETTERS of MOZART. Translated from the German Biography of Dr. LUDWIG NOHL by Lady WALLACE. 2 vols. post 8vo. with Portraits of Mozart and his Sister. [*Nearly ready*.

FELIX MENDELSSOHN'S LETTERS from ITALY and SWITZERLAND, and Letters from 1833 to 1847. Translated by Lady WALLACE. With Portrait, 2 vols. crown 8vo. 5*s*. each.

The LIFE of ROBERT FRAMPTON, D.D. Bishop of Gloucester, deprived as a Non-Juror in 1689. Edited by T. SIMPSON EVANS, M.A. Vicar of Shoreditch. Crown 8vo. Portrait, 10*s*. 6*d*.

AUTOBIOGRAPHY. By JOHN STUART MILL. 8vo. price 7*s*. 6*d*.

The LIFE of NAPOLEON III. derived from State Records, Unpublished Family Correspondence, and Personal Testimony. By BLANCHARD JERROLD. 4 vols. 8vo. with numerous Portraits and Facsimiles. VOLS. I. and II. price 18*s*. each. The Third Volume is in the press.

ESSAYS in MODERN MILITARY BIOGRAPHY. By CHARLES CORNWALLIS CHESNEY, Lieutenant-Colonel in the Royal Engineers. 8vo. 12*s*. 6*d*.

The MEMOIRS of SIR JOHN RERESBY, of Thrybergh, Bart. M.P. for York, &c. 1634—1689. Written by Himself. Edited from the Original Manuscript by JAMES J. CARTWRIGHT, M.A. 8vo. price 21*s*.

ISAAC CASAUBON, 1559–1614. By MARK PATTISON, Rector of Lincoln College, Oxford. 8vo. 18s.

LEADERS of PUBLIC OPINION in IRELAND; Swift, Flood, Grattan, and O'Connell. By W. E. H. LECKY, M.A. New Edition, revised and enlarged. Crown 8vo. price 7s. 6d.

DICTIONARY of GENERAL BIOGRAPHY; containing Concise Memoirs and Notices of the most Eminent Persons of all Countries, from the Earliest Ages. By W. L. R. CATES. Medium 8vo. price 25s.

LIFE of the DUKE of WELLINGTON. By the Rev. G. R. GLEIG, M.A. Popular Edition, carefully revised; with copious Additions. Crown 8vo. with Portrait, 5s.

MEMOIRS of SIR HENRY HAVELOCK, K.C.B. By JOHN CLARK MARSHMAN. Cabinet Edition, with Portrait. Crown 8vo. price 3s. 6d.

VICISSITUDES of FAMILIES. By Sir J. BERNARD BURKE, C.B. Ulster King of Arms. New Edition, enlarged. 2 vols. crown 8vo. 21s.

ESSAYS in ECCLESIASTICAL BIOGRAPHY. By the Right Hon. Sir J. STEPHEN, LL.D. Cabinet Edition. Crown 8vo. 7s. 6d.

MAUNDER'S BIOGRAPHICAL TREASURY. Latest Edition, reconstructed, thoroughly revised, and in great part rewritten; with 1,500 additional Memoirs and Notices, by W. L. R. CATES. Fcp. 8vo. 6s. cloth; 10s. 6d. calf.

LETTERS and LIFE of FRANCIS BACON, including all his Occasional Works. Collected and edited, with a Commentary, by J. SPEDDING, Trin. Coll. Cantab. Complete in 7 vols. 8vo. £4. 4s.

The LIFE, WORKS, and OPINIONS of HEINRICH HEINE. By WILLIAM STIGAND. 2 vols. 8vo. with Portrait of Heine, price 28s.

BIOGRAPHICAL and CRITICAL ESSAYS, reprinted from Reviews, with Additions and Corrections. Second Edition of the Second Series. By A. HAYWARD, Q.C. 2 vols. 8vo. price 28s. THIRD SERIES, in 1 vol. 8vo. price 14s.

Criticism, Philosophy, Polity, &c.

The LAW of NATIONS considered as INDEPENDENT POLITICAL COMMUNITIES; the Rights and Duties of Nations in Time of War. By Sir TRAVERS TWISS, D.C.L., F.R.S. Second Edition, revised. 8vo. 21s.

CHURCH and STATE: their relations Historically Developed. By T. HEINRICH GEFFCKEN, Professor of International Law in the University of Strasburg. Translated and edited with the Author's assistance by E. FAIRFAX TAYLOR. 2 vols. 8vo. 42s.

The INSTITUTES of JUSTINIAN; with English Introduction, Translation and Notes. By T. C. SANDARS, M.A. Sixth Edition. 8vo. 18s.

A SYSTEMATIC VIEW of the SCIENCE of JURISPRUDENCE. By SHELDON AMOS, M.A. Professor of Jurisprudence to the Inns of Court, London. 8vo. price 18s.

A PRIMER of the ENGLISH CONSTITUTION and GOVERNMENT. By SHELDON AMOS, M.A. Professor of Jurisprudence to the Inns of Court. Second Edition, revised. Crown 8vo. 6s.

A SKETCH of the HISTORY of TAXES in ENGLAND from the Earliest Times to the Present Day. By STEPHEN DOWELL. VOL. I. to the Civil War 1642. 8vo. 10*s*. 6*d*.

OUTLINES of CIVIL PROCEDURE. Being a General View of the Supreme Court of Judicature and of the whole Practice in the Common Law and Chancery Divisions under all the Statutes now in force. By EDWARD STANLEY ROSCOE, Barrister-at-Law. 12mo. price 3*s*. 6*d*.

Our NEW JUDICIAL SYSTEM and CIVIL PROCEDURE, as Reconstructed under the Judicature Acts, including the Act of 1876 ; with Comments on their Effect and Operation. By W. F. FINLASON. Crown 8vo. 10*s*. 6*d*.

SOCRATES and the SOCRATIC SCHOOLS. Translated from the German of Dr. E. ZELLER, with the Author's approval, by the Rev. OSWALD J. REICHEL, M.A. Crown 8vo. New Edition in the Press.

The STOICS, EPICUREANS, and SCEPTICS. Translated from the German of Dr. E. ZELLER, with the Author's approval, by OSWALD J. REICHEL, M.A. Crown 8vo. price 14*s*.

PLATO and the OLDER ACADEMY. Translated from the German of Dr. EDUARD ZELLER by S. FRANCES ALLEYNE and ALFRED GOODWIN, B.A. Fellow of Balliol College, Oxford. Crown 8vo. 18*s*.

The ETHICS of ARISTOTLE, with Essays and Notes. By Sir A. GRANT, Bart. M.A. LL.D. Third Edition. 2 vols. 8vo. 32*s*.

The POLITICS of ARISTOTLE; Greek Text, with English Notes. By RICHARD CONGREVE, M.A. 8vo. 18*s*.

ARISTOTLE'S POLITICS. Books I. III. IV. (VII.) The Greek Text of Bekker, with an English Translation by W. E. BOLLAND, M.A. and Short Introductory Essays by A. LANG, M.A. Crown 8vo. 7*s*. 6*d*.

The NICOMACHEAN ETHICS of ARISTOTLE newly translated into English. By R. WILLIAMS, B.A. Fellow and late Lecturer of Merton College, and sometime Student of Christ Church, Oxford. 8vo. 7*s*. 6*d*.

ELEMENTS of LOGIC. By R. WHATELY, D.D. sometime Archbishop of Dublin. 8vo. 10*s*. 6*d*. Crown 8vo. 4*s*. 6*d*.

LOGIC, DEDUCTIVE and INDUCTIVE. By ALEXANDER BAIN, LL.D. In Two PARTS, crown 8vo. 10*s*. 6*d*. Each Part may be had separately :—
PART I. *Deduction,* 4*s*. PART II. *Induction,* 6*s*. 6*d*.

PICTURE LOGIC; an Attempt to Popularise the Science of Reasoning by the combination of Humorous Pictures with Examples of Reasoning taken from Daily Life. By A. SWINBOURNE, B.A. Fcp. 8vo. Woodcuts, 5*s*.

ELEMENTS of RHETORIC. By R. WHATELY, D.D. sometime Archbishop of Dublin. 8vo. 10*s*. 6*d*. Crown 8vo. 4*s*. 6*d*.

COMTE'S SYSTEM of POSITIVE POLITY, or TREATISE upon SOCIOLOGY. Translated from the Paris Edition of 1851-1854, and furnished with Analytical Tables of Contents. In Four Volumes, 8vo. each forming in some degree an independent Treatise :—

VOL. I. General View of Positivism and its Introductory Principles. Translated by J. H. BRIDGES, M.B. Price 21*s*.

VOL. II. The Social Statics, or the Abstract Laws of Human Order. Translated by F. HARRISON, M.A. Price 14*s*.

VOL. III. The Social Dynamics, or the General Laws of Human Progress (the Philosophy of History). Translated by Professor E. S. BEESLY, M.A. 8vo. 21*s*.

VOL. IV. The Synthesis of the Future of Mankind. Translated by R. CONGREVE, M.D. ; and an Appendix, containing Comte's Early Essays, translated by H. D. Hutton, B.A.   [*In the press.*

DEMOCRACY in AMERICA. By Alexis de Tocqueville. Translated by Henry Reeve, Esq. 2 vols. crown 8vo. 16*s*.

On the INFLUENCE of AUTHORITY in MATTERS of OPINION. By the late Sir George Cornewall Lewis, Bart. 8vo. 14*s*.

BACON'S ESSAYS with ANNOTATIONS. By R. Whately, D.D. late Archbishop of Dublin. Fourth Edition. 8vo. price 10*s*. 6*d*.

LORD BACON'S WORKS, collected and edited by J. Spedding, M.A. R. L. Ellis, M.A. and D. D. Heath. 7 vols. 8vo. price £3. 13*s*. 6*d*.

On REPRESENTATIVE GOVERNMENT. By John Stuart Mill. Crown 8vo. price 2*s*.

On LIBERTY. By John Stuart Mill. Post 8vo. 7*s*. 6*d*. Crown 8vo. price 1*s*. 4*d*.

PRINCIPLES of POLITICAL ECONOMY. By John Stuart Mill. 2 vols. 8vo. 30*s*. Or in 1 vol. crown 8vo. price 5*s*.

ESSAYS on SOME UNSETTLED QUESTIONS of POLITICAL ECONOMY. By John Stuart Mill. 8vo. 6*s*. 6*d*.

UTILITARIANISM. By John Stuart Mill. 8vo. 5*s*.

DISSERTATIONS and DISCUSSIONS; Political, Philosophical, and Historical. By John Stuart Mill. 4 vols. 8vo. price £2. 6*s*. 6*d*.

EXAMINATION of Sir. W. HAMILTON'S PHILOSOPHY, and of the Principal Philosophical Questions discussed in his Writings. By John Stuart Mill. 8vo. 16*s*.

A SYSTEM of LOGIC, RATIOCINATIVE and INDUCTIVE. By John Stuart Mill. Two vols. 8vo. 25*s*.

An OUTLINE of the NECESSARY LAWS of THOUGHT; a Treatise on Pure and Applied Logic. By the Most Rev. W. Thomson, Lord Archbishop of York, D.D. F.R.S. Crown 8vo. price 6*s*.

PRINCIPLES of ECONOMICAL PHILOSOPHY. By Henry Dunning Macleod, M.A. Barrister-at-Law. Second Edition. In Two Volumes. Vol. I. 8vo. price 15*s*. Vol. II. Part I. price 12*s*. Vol. II. Part II. *just ready*.

SPEECHES of the RIGHT HON. LORD MACAULAY, corrected by Himself. Crown 8vo. 3*s*. 6*d*.

FAMILIES of SPEECH; Four Lectures delivered before the Royal Institution. By the Rev. Canon Farrar, F.R.S. Crown 8vo. 3*s*. 6*d*.

CHAPTERS on LANGUAGE. By the Rev. Canon Farrar, F.R.S. Crown 8vo. 5*s*.

HANDBOOK of the ENGLISH LANGUAGE. For the use of Students of the Universities and the Higher Classes in Schools. By R. G. Latham, M.A. M.D. Crown 8vo. price 6*s*.

DICTIONARY of the ENGLISH LANGUAGE. By R. G. Latham, M.A. M.D. Abridged from Dr. Latham's Edition of Johnson's English Dictionary, and condensed into One Volume. Medium 8vo. price 24*s*.

A DICTIONARY of the ENGLISH LANGUAGE. By R. G. Latham, M.A. M.D. Founded on the Dictionary of Dr. Samuel Johnson, as edited by the Rev. H. J. Todd, with numerous Emendations and Additions. In Four Volumes, 4to. price £7.

ENGLISH SYNONYMES. By E. JANE WHATELY. Edited by Archbishop WHATELY. Fifth Edition. Fcp. 8vo. price 3s.

THESAURUS of ENGLISH WORDS and PHRASES, classified and arranged so as to facilitate the Expression of Ideas, and assist in Literary Composition. By P. M. ROGET, M.D. Crown 8vo. 10s. 6d.

LECTURES on the SCIENCE of LANGUAGE. By F. MAX MÜLLER, M.A. &c. The Eighth Edition. 2 vols. crown 8vo. 16s.

MANUAL of ENGLISH LITERATURE, Historical and Critical. By THOMAS ARNOLD, M.A. Crown 8vo. 7s. 6d.

HISTORICAL and CRITICAL COMMENTARY on the OLD TESTA-MENT; with a New Translation. By M. M. KALISCH, Ph.D. VOL. I. *Genesis*, 8vo. 18s. or adapted for the General Reader, 12s. VOL. II. *Exodus*, 15s. or adapted for the General Reader, 12s. VOL. III. *Leviticus*, PART I. 15s. or adapted for the General Reader, 8s. VOL. IV. *Leviticus*, PART II. 15s. or adapted for the General Reader, 8s.

A DICTIONARY of ROMAN and GREEK ANTIQUITIES, with about Two Thousand Engravings on Wood from Ancient Originals, illustrative of the Industrial Arts and Social Life of the Greeks and Romans. By A. RICH, B.A. Third Edition, revised and improved. Crown 8vo. price 7s. 6d.

A LATIN-ENGLISH DICTIONARY. By JOHN T. WHITE, D.D. Oxon. and J. E. RIDDLE, M.A. Oxon. 1 vol. 4to. 28s.

WHITE'S COLLEGE LATIN-ENGLISH DICTIONARY (Intermediate Size), abridged for the use of University Students from the Parent Work (as above). Medium 8vo. 15s.

WHITE'S JUNIOR STUDENT'S COMPLETE LATIN-ENGLISH and ENGLISH-LATIN DICTIONARY. Square 12mo. price 12s.

  Separately { The ENGLISH-LATIN DICTIONARY, price 5s. 6d.
        { The LATIN-ENGLISH DICTIONARY, price 7s. 6d.

A LATIN-ENGLISH DICTIONARY, adapted for the Use of Middle-Class Schools. By JOHN T. WHITE, D.D. Oxon. Square fcp. 8vo. price 3s.

An ENGLISH-GREEK LEXICON, containing all the Greek Words used by Writers of good authority. By C. D. YONGE, M.A. 4to. price 21s.

Mr. YONGE'S NEW LEXICON, English and Greek, abridged from his larger work (as above). Square 12mo. price 8s. 6d.

LIDDELL and SCOTT'S GREEK-ENGLISH LEXICON. Sixth Edition. Crown 4to. price 36s.

A LEXICON, GREEK and ENGLISH, abridged from LIDDELL and SCOTT's *Greek-English Lexicon.* Fourteenth Edition. Square 12mo. 7s. 6d.

A PRACTICAL DICTIONARY of the FRENCH and ENGLISH LAN-GUAGES. By L. CONTANSEAU. Post 8vo. 7s. 6d.

CONTANSEAU'S POCKET DICTIONARY, French and English, abridged from the above by the Author. Square 18mo. 3s. 6d.

A NEW POCKET DICTIONARY of the GERMAN and ENGLISH LANGUAGES. By F. W. LONGMAN, Balliol College, Oxford. 18mo. 5s.

NEW PRACTICAL DICTIONARY of the GERMAN LANGUAGE; German-English and English-German. By the Rev. W. L. BLACKLEY, M.A. and Dr. CARL MARTIN FRIEDLÄNDER. Post 8vo. 7s. 6d.

Miscellaneous Works and *Popular Metaphysics.*

GERMAN HOME LIFE. Reprinted, with Revision and Additions, from *Fraser's Magazine.* Third Edition. Crown 8vo. 6*s.*

THE MISCELLANEOUS WORKS of THOMAS ARNOLD, D.D. Late Head Master of Rugby School. 8vo. 7*s.* 6*d.*

MISCELLANEOUS and POSTHUMOUS WORKS of the Late HENRY THOMAS BUCKLE. Edited, with a Biographical Notice, by HELEN TAYLOR. 3 vols. 8vo. price 52*s.* 6*d.*

MISCELLANEOUS WRITINGS of JOHN CONINGTON, M.A. late Corpus Professor of Latin in the University of Oxford. Edited by J. A. SYMONDS, M.A. With a Memoir by H. J. S. SMITH, M.A. 2 vols. 8vo. 28*s.*

ESSAYS, CRITICAL and BIOGRAPHICAL. Contributed to the *Edinburgh Review.* By HENRY ROGERS. 2 vols. crown 8vo. price 12*s.*

ESSAYS on some THEOLOGICAL CONTROVERSIES of the TIME. Contributed chiefly to the *Edinburgh Review.* By HENRY ROGERS. New Edition, with Additions. Crown 8vo. price 6*s.*

The ESSAYS and CONTRIBUTIONS of A. K. H. B. Uniform Cabinet Edition, in crown 8vo. :—

> Recreations of a Country Parson. Two Series, 3*s.* 6*d.* each.
> The Common-place Philosopher in Town and Country. 3*s.* 6*d.*
> Leisure Hours in Town. 3*s.* 6*d.*
> The Autumn Holidays of a Country Parson. 3*s.* 6*d.*
> Seaside Musings on Sundays and Week-Days. 3*s.* 6*d.*
> The Graver Thoughts of a Country Parson. Three Series, 3*s.* 6*d.* each.
> Critical Essays of a Country Parson. 3*s.* 6*d.*,
> Sunday Afternoons in the Parish Church of a University City. 3*s.* 6*d.*
> Lessons of Middle Age. 3*s.* 6*d.*
> Counsel and Comfort spoken from a City Pulpit. 3*s.* 6*d.*
> Changed Aspects of Unchanged Truths. 3*s.* 6*d.*
> Present-day Thoughts. 3*s.* 6*d.*
> Landscapes, Churches, and Moralities. 3*s.* 6*d.*

SHORT STUDIES on GREAT SUBJECTS. By JAMES ANTHONY FROUDE, M.A. late Fellow of Exeter Coll. Oxford. 2 vols. crown 8vo. price 12*s.* or 2 vols. demy 8vo. price 24*s.* Third Series in the press.

SELECTIONS from the WRITINGS of LORD MACAULAY. Edited, with Occasional Explanatory Notes, by GEORGE OTTO TREVELYAN, M.P. Crown 8vo. price 6*s.*

LORD MACAULAY'S MISCELLANEOUS WRITINGS :—
> LIBRARY EDITION. 2 vols. 8vo. Portrait, 21*s.*
> PEOPLE's EDITION. 1 vol. crown 8vo. 4*s.* 6*d.*

LORD MACAULAY'S MISCELLANEOUS WRITINGS and SPEECHES. STUDENT's EDITION, in crown 8vo. price 6*s.*

The Rev. SYDNEY SMITH'S MISCELLANEOUS WORKS; including his Contributions to the *Edinburgh Review.* Crown 8vo. 6*s.*

The WIT and WISDOM of the Rev. SYDNEY SMITH; a Selection of the most memorable Passages in his Writings and Conversation. 16mo. 3*s.* 6*d.*

The ECLIPSE of FAITH; or, a Visit to a Religious Sceptic. By HENRY ROGERS. Latest Edition. Fcp. 8vo. price 5*s.*

DEFENCE of the ECLIPSE of FAITH, by its Author; a rejoinder to Dr. Newman's *Reply.* Latest Edition. Fcp 8vo. price 3*s.* 6*d.*

CHIPS from a GERMAN WORKSHOP; Essays on the Science of Religion, on Mythology, Traditions, and Customs, and on the Science of Language. By F. MAX MÜLLER, M.A. &c. 4 vols. 8vo. £2. 18*s*.

ANALYSIS of the PHENOMENA of the HUMAN MIND. By JAMES MILL. A New Edition, with Notes, Illustrative and Critical, by ALEXANDER BAIN, ANDREW FINDLATER, and GEORGE GROTE. Edited, with additional Notes, by JOHN STUART MILL. 2 vols. 8vo. price 28*s*.

An INTRODUCTION to MENTAL PHILOSOPHY, on the Inductive Method. By J. D. MORELL, M.A. LL.D. 8vo. 12*s*.

PHILOSOPHY WITHOUT ASSUMPTIONS. By the Rev. T. P. KIRKMAN, F.R.S. Rector of Croft, near Warrington. 8vo. 10*s*. 6*d*.

The SENSES and the INTELLECT. By ALEXANDER BAIN, M.D. Professor of Logic in the University of Aberdeen. Third Edition. 8vo. 15*s*.

The EMOTIONS and the WILL. By ALEXANDER BAIN, LL.D. Professor of Logic in the University of Aberdeen. Third Edition, thoroughly revised, and in great part re-written. 8vo. price 15*s*.

MENTAL and MORAL SCIENCE: a Compendium of Psychology and Ethics. By the same Author. Third Edition. Crown 8vo. 10*s*. 6*d*. Or separately: PART I. *Mental Science*, 6*s*. 6*d*. PART II. *Moral Science*, 4*s*. 6*d*.

APPARITIONS; a Narrative of Facts. By the Rev. B. W. SAVILE, M.A. Author of ' The Truth of the Bible ' &c. Crown 8vo. price 4*s*. 6*d*.

HUME'S TREATISE of HUMAN NATURE, Edited, with Notes &c. by T. H. GREEN, Fellow and Tutor, Ball. Coll. and T. H. GROSE, Fellow and Tutor, Queen's Coll. Oxford. 2 vols. 8vo. 28*s*.

ESSAYS MORAL, POLITICAL, and LITERARY. By DAVID HUME. By the same Editors. 2 vols. 8vo. price 28*s*.

   *** The above form a complete and uniform Edition of DAVID HUME'S Philosophical Works.

The PHILOSOPHY of NECESSITY; or, Natural Law as applicable to Mental, Moral, and Social Science. By CHARLES BRAY. 8vo. 9*s*.

Astronomy, Meteorology, Popular Geography, &c.

BRINKLEY'S ASTRONOMY. Revised and partly re-written by J. W. STUBBS, D.D. Fellow and Tutor of Trinity College, Dublin, and F. BRÜNNOW, Ph.D. Astronomer Royal of Ireland. Crown 8vo. 6*s*.

OUTLINES of ASTRONOMY. By Sir J. F. W. HERSCHEL, Bart. M.A. Latest Edition, with Plates and Diagrams. Square crown 8vo. 12*s*.

ESSAYS on ASTRONOMY, a Series of Papers on Planets and Meteors, the Sun and Sun-surrounding Space, Stars and Star-Cloudlets; with a Dissertation on the Transit of Venus. By R. A. PROCTOR, B.A. With Plates and Woodcuts. 8vo. 12*s*.

THE TRANSITS of VENUS; a Popular Account of Past and Coming Transits, from the first observed by Horrocks A.D. 1639 to the Transit of A.D. 2012. By R. A. PROCTOR, B.A. Second Edition, with 20 Plates (12 coloured) and 38 Woodcuts. Crown 8vo. 8*s*. 6*d*.

The UNIVERSE and the COMING TRANSITS : Presenting Researches into and New Views respecting the Constitution of the Heavens; together with an Investigation of the Conditions of the Coming Transits of Venus. By R. A. PROCTOR, B.A. With 22 Charts and 22 Woodcuts. 8vo. 16s.

The MOON; her Motions, Aspect, Scenery, and Physical Condition. By R. A. PROCTOR, B.A. With Plates, Charts, Woodcuts, and Three Lunar Photographs. Crown 8vo. 15s.

The SUN; RULER, LIGHT, FIRE, and LIFE of the PLANETARY SYSTEM. By R. A. PROCTOR, B.A. Third Edition, with 10 Plates (7 coloured) and 107 Figures on Wood. Crown 8vo. 14s.

OTHER WORLDS THAN OURS; the Plurality of Worlds Studied under the Light of Recent Scientific Researches. By R. A. PROCTOR, B.A. Third Edition, with 14 Illustrations. Crown 8vo. 10s. 6d.

The ORBS AROUND US; Familiar Essays on the Moon and Planets, Meteors and Comets, the Sun and Coloured Pairs of Stars. By R. A. PROCTOR, B.A. Second Edition, with Charts and 4 Diagrams. Crown 8vo. price 7s. 6d.

SATURN and its SYSTEM. By R. A. PROCTOR, B.A. 8vo. with 14 Plates, 14s.

A NEW STAR ATLAS, for the Library, the School, and the Observatory, in Twelve Circular Maps (with Two Index Plates). Intended as a Companion to 'Webb's Celestial Objects for Common Telescopes.' With a Letterpress Introduction on the Study of the Stars, illustrated by 9 Diagrams. By R. A. PROCTOR, B.A. Crown 8vo. 5s.

CELESTIAL OBJECTS for COMMON TELESCOPES. By the Rev. T. W. WEBB, M.A. F.R.A.S. Third Edition, revised and enlarged; with Maps, Plate, and Woodcuts. Crown 8vo. price 7s. 6d.

The MOON, and the Condition and Configurations of its Surface. By EDMUND NEISON, Fellow of the Royal Astronomical Society, &c. With 26 Maps and 5 Plates. Medium 8vo. 31s. 6d.

SCHELLEN'S SPECTRUM ANALYSIS, in its application to Terrestrial Substances and the Physical Constitution of the Heavenly Bodies. Translated by JANE and C. LASSELL; edited, with Notes, by W. HUGGINS, LL.D. F.R.S. With 13 Plates (6 coloured) and 223 Woodcuts. 8vo. price 28s.

AIR and RAIN; the Beginnings of a Chemical Climatology. By ROBERT ANGUS SMITH, Ph.D. F.R.S. F.C.S. With 8 Illustrations. 8vo. 24s.

AIR and its RELATIONS to LIFE. By W. N. HARTLEY, F.C.S. Demonstrator of Chemistry at King's College, London. Second Edition, with 66 Woodcuts. Small 8vo. 6s.

DOVE'S LAW of STORMS, considered in connexion with the Ordinary Movements of the Atmosphere. Translated by R. H. SCOTT, M.A. 8vo. 10s. 6d.

The PUBLIC SCHOOLS ATLAS of MODERN GEOGRAPHY. In 31 entirely new Coloured Maps. Edited, with an Introduction, by the Rev. G. BUTLER, M.A. Imperial 8vo. or imperial 4to. 5s. cloth.

The PUBLIC SCHOOLS ATLAS of ANCIENT GEOGRAPHY, in 28 entirely new Coloured Maps. Edited by the Rev. G. BUTLER, M.A. Imperial 8vo. or imperial 4to. 7s. 6d. cloth.

KEITH JOHNSTON'S GENERAL DICTIONARY of GEOGRAPHY,
Descriptive, Physical, Statistical, and Historical; forming a complete Gazetteer
of the World. New Edition (1877), revised and corrected. 8vo. price 42*s*.

MAUNDER'S TREASURY of GEOGRAPHY, Physical, Historical,
Descriptive, and Political. Edited by W. HUGHES, F.R.G.S. Revised Edition,
with 7 Maps and 16 Plates. Fcp. 6*s*. cloth, or 10*s*. 6*d*. bound in calf.

Natural History and *Popular Science.*

TEXT-BOOKS of SCIENCE, MECHANICAL and PHYSICAL,
adapted for the use of Artisans and of Students in Public and Science Schools.

The following Text-Books in this Series may now be had:—
> ANDERSON's Strength of Materials, small 8vo. 3*s*. 6*d*.
> ARMSTRONG's Organic Chemistry, 3*s*. 6*d*.
> BARRY's Railway Appliances, 3s. 6d.
> BLOXAM's Metals, 3*s*. 6d.
> GOODEVE's Elements of Mechanism, 3*s*. 6*d*.
> ———— Principles of Mechanics, 3*s*. 6*d*.
> GRIFFIN's Algebra and Trigonometry, 3*s*. 6*d*.
> JENKIN's Electricity and Magnetism, 3*s*. 6*d*.
> MAXWELL's Theory of Heat, 3*s*. 6*d*.
> MERRIFIELD's Technical Arithmetic and Mensuration, 3*s*. 6*d*.
> MILLER's Inorganic Chemistry. 3*s*. 6*d*.
> PREECE & SIVEWRIGHT's Telegraphy, 3*s*. 6*d*.
> SHELLEY's Workshop Appliances, 3*s*. 6*d*.
> THOMÉ's Structural and Physiological Botany, 6*s*.
> THORPE's Quantitative Chemical Analysis, 4*s*. 6*d*.
> THORPE & MUIR's Qualitative Analysis, 3*s*. 6*d*.
> TILDEN's Chemical Philosophy, 3*s*. 6*d*.
> UNWIN's Machine Design, 3*s*. 6*d*.
> WATSON's Plane and Solid Geometry, 3*s*. 6*d*.

*** Other Text-Books in continuation of this Series are in active preparation.

ELEMENTARY TREATISE on PHYSICS, Experimental and Applied.
Translated and edited from GANOT's *Éléments de Physique* by E. ATKINSON,
Ph.D. F.C.S. Seventh Edition, revised and enlarged; with 4 Coloured Plates
and 758 Woodcuts. Post 8vo. 15*s*.

NATURAL PHILOSOPHY for GENERAL READERS and YOUNG
PERSONS; being a Course of Physics divested of Mathematical Formulæ
expressed in the language of daily life. Translated from GANOT's *Cours de
Physique* and by E. ATKINSON, Ph.D. F.C.S. Second Edition, with 2 Plates
and 429 Woodcuts. Crown 8vo. price 7*s*. 6*d*.

ARNOTT'S ELEMENTS of PHYSICS or NATURAL PHILOSOPHY.
Seventh Edition, edited by A. BAIN, LL.D. and A. S. TAYLOR, M.D. F.R.S.
Crown 8vo. Woodcuts, 12*s*. 6*d*.

HELMHOLTZ'S POPULAR LECTURES on SCIENTIFIC SUBJECTS.
Translated by E. ATKINSON, Ph.D. F.C.S. Professor of Experimental Science,
Staff College. With an Introduction by Professor TYNDALL. 8vo. with nume-
rous Woodcuts, price 12*s*. 6*d*.

On the SENSATIONS of TONE as a Physiological Basis for the
Theory of Music. By HERMANN L. F. HELMHOLTZ, M.D. Professor of Physics
in the University of Berlin. Translated, with the Author's sanction, from the
Third German Edit'on, with Additional Notes and an Additional Appendix, by
ALEXANDER J. ELLIS, F.R.S. &c. 8vo. price 36*s*.

The **HISTORY of MODERN MUSIC**, a Course of Lectures delivered at the Royal Institution of Great Britain. By JOHN HULLAH, Professor of Vocal Music in Queen's College and Bedford College, and Organist of Charterhouse. 8vo. 8s. 6d.

The **TRANSITION PERIOD of MUSICAL HISTORY**; a Second Course of Lectures on the History of Music from the Beginning of the Seventeenth to the Middle of the Eighteenth Century, delivered at the Royal Institution. By JOHN HULLAH. 8vo. 10s. 6d.

SOUND. By JOHN TYNDALL, LL.D. D.C.L. F.R.S. Third Edition, including Recent Researches on Fog-Signalling; Portrait and Woodcuts. Crown 8vo. 10s. 6d.

HEAT a MODE of MOTION. By JOHN TYNDALL, LL.D. D.C.L. F.R.S. Fifth Edition. Plate and Woodcuts. Crown 8vo. 10s. 6d.

CONTRIBUTIONS to MOLECULAR PHYSICS in the DOMAIN of RADIANT HEAT. By J. TYNDALL, LL.D. D.C.L. F.R.S. With 2 Plates and 31 Woodcuts. 8vo. 16s.

RESEARCHES on DIAMAGNETISM and MAGNE-CRYSTALLIC ACTION; including the Question of Diamagnetic Polarity. By J. TYNDALL M.D. D.C.L. F.R.S. With 6 plates and many Woodcuts. 8vo. 14s.

NOTES of a COURSE of SEVEN LECTURES on ELECTRICAL PHENOMENA and THEORIES, delivered at the Royal Institution, A.D. 1870. By JOHN TYNDALL, LL.D., D.C.L., F.R.S. Crown 8vo. 1s. sewed; 1s. 6d. cloth.

SIX LECTURES on LIGHT delivered in America in 1872 and 1873. By JOHN TYNDALL, LL.D. D.C.L. F.R.S. Second Edition, with Portrait, Plate, and 59 Diagrams. Crown 8vo. 7s. 6d.

NOTES of a COURSE of NINE LECTURES on LIGHT delivered at the Royal Institution, A.D. 1869. By JOHN TYNDALL, LL.D. D.C.L. F.R.S. Crown 8vo. price 1s. sewed, or 1s. 6d. cloth.

FRAGMENTS of SCIENCE. By JOHN TYNDALL, LL.D. D.C.L. F.R.S. Third Edition, with a New Introduction. Crown 8vo. 10s. 6d.

LIGHT SCIENCE for LEISURE HOURS; a Series of Familiar Essays on Scientific Subjects, Natural Phenomena, &c. By R. A. PROCTOR, B.A. First and Second Series. Crown 8vo. 7s. 6d. each.

A TREATISE on MAGNETISM, General and Terrestrial. By HUMPHREY LLOYD, D.D. D.C.L., Provost of Trinity College, Dublin. 8vo. 10s. 6d.

ELEMENTARY TREATISE on the WAVE-THEORY of LIGHT. By HUMPHREY LLOYD, D.D. D.C.L. Provost of Trinity College, Dublin. Third Edition, revised and enlarged. 8vo. price 10s. 6d.

The **CORRELATION of PHYSICAL FORCES**. By the Hon. Sir W. R. GROVE, M.A. F.R.S. one of the Judges of the Court of Common Pleas. Sixth Edition, with other Contributions to Science. 8vo. price 15s.

The **COMPARATIVE ANATOMY and PHYSIOLOGY of the VERTEBRATE ANIMALS**. By RICHARD OWEN, F.R.S. D.C.L. With 1,472 Woodcuts. 3 vols. 8vo. £3. 13s. 6d.

PRINCIPLES of ANIMAL MECHANICS. By the Rev. S. HAUGHTON, F.R.S. Fellow of Trin. Coll. Dubl. M.D. Dubl. and D.C.L. Oxon. Second Edition, with 111 Figures on Wood. 8vo. 21*s.*

ROCKS CLASSIFIED and DESCRIBED. By BERNHARD VON COTTA. English Edition, by P. H. LAWRENCE; with English, German, and French Synonymes. Post 8vo. 14*s.*

The **ANCIENT STONE IMPLEMENTS, WEAPONS, and ORNAMENTS of GREAT BRITAIN.** By JOHN EVANS, F.R.S. F.S.A. With 2 Plates and 476 Woodcuts. 8vo. price 28*s.*

The **GEOLOGY of ENGLAND and WALES;** A Concise Account of the Lithological Characters, Leading Fossils, and Economic Products of the Rocks; with Notes on the Physical Features of the Country. By H. B. WOODWARD, F.G.S. With a Coloured Map and 29 Woodcuts. Crown 8vo. 14*s.*

The **PRIMÆVAL WORLD of SWITZERLAND.** By Professor OSWALD HEER, of the University of Zurich. Edited by JAMES HEYWOOD, M.A. F.R.S. President of the Statistical Society. With a Coloured Map, 19 Plates in Lithography and Chromoxylography, and 372 Woodcuts. 2 vols. 8vo. 28*s.*

The **PUZZLE of LIFE and HOW it HAS BEEN PUT TOGETHER:** a Short History of Vegetable and Animal Life upon the Earth from the Earliest Times; including an Account of Pre-Historic Man, his Weapons, Tools, and Works. By A. NICOLS, F.R.G.S. With 12 Illustrations. Crown 8vo. 5*s.*

The **ORIGIN of CIVILISATION and the PRIMITIVE CONDITION** of MAN; Mental and Social Condition of Savages. By Sir JOHN LUBBOCK, Bart. M.P. F.R.S. Third Edition, with 25 Woodcuts. 8vo. 18*s.*

BIBLE ANIMALS; being a Description of every Living Creature mentioned in the Scriptures, from the Ape to the Coral. By the Rev. J. G. WOOD, M.A. F.L.S. With about 112 Vignettes on Wood. 8vo. 14*s.*

HOMES WITHOUT HANDS; a Description of the Habitations of Animals, classed according to their Principle of Construction. By the Rev. J. G. WOOD, M.A. F.L.S. With about 140 Vignettes on Wood. 8vo. 14*s.*

INSECTS AT HOME; a Popular Account of British Insects, their Structure, Habits, and Transformations. By the Rev. J. G. WOOD, M.A. F.L.S. With upwards of 700 Illustrations. 8vo. price 14*s.*

INSECTS ABROAD; a Popular Account of Foreign Insects, their Structure, Habits, and Transformations. By J. G. WOOD, M.A. F.L.S. Printed and illustrated uniformly with 'Insects at Home.' 8vo. price 14*s.*

STRANGE DWELLINGS; a description of the Habitations of Animals, abridged from 'Homes without Hands.' By the Rev. J. G. WOOD, M.A. F.L.S. With about 60 Woodcut Illustrations. Crown 8vo. price 7*s.* 6*d.*

OUT of DOORS; a Selection of original Articles on Practical Natural History. By the Rev. J. G. WOOD, M.A. F.L.S. With Eleven Illustrations from Original Designs engraved on Wood by G. Pearson. Crown 8vo. price 7*s.* 6*d.*

A FAMILIAR HISTORY of BIRDS. By E. STANLEY, D.D. F.R.S. late Lord Bishop of Norwich. Seventh Edition, with Woodcuts. Fcp. 3*s.* 6*d.*

KIRBY and SPENCE'S INTRODUCTION to ENTOMOLOGY, or Elements of the Natural History of Insects. 7th Edition. Crown 8vo. 5*s.*

The SEA and its LIVING WONDERS. By Dr. GEORGE HARTWIG.
Latest revised Edition. 8vo. with many Illustrations, 10s. 6d.

The TROPICAL WORLD. By Dr. GEORGE HARTWIG. With above 160
Illustrations. Latest revised Edition. 8vo. price 10s. 6d.

The SUBTERRANEAN WORLD. By Dr. GEORGE HARTWIG. With
3 Maps and about 80 Woodcuts, including 8 full size of page. 8vo. price 10s. 6d.

The POLAR WORLD, a Popular Description of Man and Nature in the
Arctic and Antarctic Regions of the Globe. By Dr. GEORGE HARTWIG. With
8 Chromoxylographs, 3 Maps, and 85 Woodcuts. 8vo. 10s. 6d.

THE AERIAL WORLD. By Dr. G. HARTWIG. New Edition, with 8
Chromoxylographs and 60 Woodcut Illustrations. 8vo. price 21s.

MAUNDER'S TREASURY of NATURAL HISTORY, or Popular
Dictionary of Birds, Beasts, Fishes, Reptiles, Insects, and Creeping Things.
With above 900 Woodcuts. Fcp. 8vo. price 6s. cloth, or 10s. 6d. bound in calf.

MAUNDER'S SCIENTIFIC and LITERARY TREASURY. New
Edition, thoroughly revised and in great part rewritten, with above 1,000
new Articles, by J. Y. JOHNSON. Fcp. 8vo. 6s. cloth, or 10s. 6d. calf.

BRANDE'S DICTIONARY of SCIENCE, LITERATURE, and ART.
Re-edited by the Rev. GEORGE W. COX, M.A. late Scholar of Trinity College,
Oxford; assisted by Contributors of eminent Scientific and Literary Acquire-
ments. New Edition, revised. 3 vols. medium 8vo. 63s.

HANDBOOK of HARDY TREES, SHRUBS, and HERBACEOUS
PLANTS, containing Descriptions, Native Countries, &c. of a Selection of the
Best Species in Cultivation; together with Cultural Details, Comparative
Hardiness, Suitability for Particular Positions, &c. By W. B. HEMSLEY. With
264 Original Woodcuts. Medium 8vo. 12s.

DECAISNE and LE MAOUT'S GENERAL SYSTEM of BOTANY,
DESCRIPTIVE and ANALYTICAL. Translated by Mrs. HOOKER. The
Orders arranged after the Method followed in the Universities and Schools of
Great Britain; with an Appendix on the Natural Method, and other Additions,
by J. D. HOOKER, F.R.S. &c. Second Thousand, with 5,500 Woodcuts. Imperial
8vo. 31s. 6d.

THOMÉ'S TEXT-BOOK of STRUCTURAL and PHYSIOIOGICAL
BOTANY. Translated by A. W. BENNETT, M.A. With Coloured Map and
600 Woodcuts. Small 8vo. 6s.

The TREASURY of BOTANY, or Popular Dictionary of the Vegetable
Kingdom; including a Glossary of Botanical Terms. Edited by J. LINDLEY,
F.R.S. and T. MOORE, F.L.S. assisted by eminent Contributors. With 274
Woodcuts and 20 Steel Plates. Two Parts, fcp. 8vo. 12s. cloth, or 21s. calf.

The ELEMENTS of BOTANY for FAMILIES and SCHOOLS.
Tenth Edition, revised by THOMAS MOORE, F.L.S. Fcp. 8vo. with 154 Woodcuts, 2s. 6d.

The ROSE AMATEUR'S GUIDE. By THOMAS RIVERS. Fourteenth
Edition. Fcp. 8vo. 4s.

LOUDON'S ENCYCLOPÆDIA of PLANTS; comprising the Specific
Character, Description, Culture, History, &c. of all the Plants found in
Great Britain. With upwards of 12,000 Woodcuts. 8vo. 42s.

Chemistry and *Physiology.*

INTRODUCTION to the STUDY of CHEMICAL PHILOSOPHY; the Principles of Theoretical and Systematic Chemistry. By WILLIAM A. TILDEN, D.Sc. Lond. F.C.S. Lecturer on Chemistry in Clifton College. With 5 Woodcuts. Small 8vo. 3*s.* 6*d.*

A DICTIONARY of CHEMISTRY and the Allied Branches of other Sciences. By HENRY WATTS, F.R.S. assisted by eminent Contributors. Seven Volumes, medium 8vo. price £10. 16*s.* 6*d.*

SUPPLEMENTARY VOLUME, completing the Record of Chemical Discovery to the year 1876.   [*In preparation.*

ELEMENTS of CHEMISTRY, Theoretical and Practical. By W. ALLEN MILLER, M.D. late Prof. of Chemistry, King's Coll. London. New Edition. 3 vols. 8vo. PART I. CHEMICAL PHYSICS, 15*s.* PART II. INORGANIC CHEMISTRY, 21*s.* PART III. ORGANIC CHEMISTRY, New Edition in the press.

SELECT METHODS in CHEMICAL ANALYSIS, chiefly INOR-GANIC. By WILLIAM CROOKES, F.R.S. With 22 Woodcuts. Crown 8vo. price 12*s.* 6*d.*

A PRACTICAL HANDBOOK of DYEING and CALICO PRINTING. By WILLIAM CROOKES. F.R.S. With 11 Page Plates, 49 Specimens of Dyed and Printed Fabrics, and 36 Woodcuts. 8vo. 42*s.*

ANTHRACEN; its Constitution, Properties, Manufacture, and Deriva-tives, including Artificial Alizarin, Anthrapurpurin, &c. with their Applica-tions in Dyeing and Printing. By G. AUERBACH. Translated by W. CROOKES, F.R.S. 8vo. 12*s.*

OUTLINES of PHYSIOLOGY, Human and Comparative. By JOHN MARSHALL, F.R.C.S. Surgeon to the University College Hospital. 2 vols. crown 8vo. with 122 Woodcuts, 32*s.*

HEALTH in the HOUSE; a Series of Lectures on Elementary Physi-ology in its application to the Daily Wants of Man and Animals, delivered to the Wives and Children of Working Men in Leeds and Saltaire. By CATHERINE M. BUCKTON. New Edition, revised. Small 8vo. Woodcuts, 2*s.*

The Fine Arts, and *Illustrated Editions.*

A DICTIONARY of ARTISTS of the ENGLISH SCHOOL: Painters, Sculptors, Architects, Engravers, and Ornamentists; with Notices of their Lives and Works. By S. REDGRAVE. 8vo. 16*s.*

MOORE'S LALLA ROOKH, an Oriental Romance, TENNIEL's Edition, with 68 Illustrations from Original Drawings, engraved on Wood by G. Pearson and other Artists. Fcp. 4to. 21*s.*

MOORE'S IRISH MELODIES, with 161 Steel Plates from Original Drawings by D. MACLISE, R.A. and the whole of the Text engraved on the same Plates. Super-royal 8vo. 21*s.*

LORD MACAULAY'S LAYS of ANCIENT ROME. With Ninety
Original Illustrations engraved on Wood, chiefly after the Antique, from Drawings by G. SCHARF. Fcp. 4to. 21s.

MINIATURE EDITION of LORD MACAULAY'S LAYS of ANCIENT
ROME, with the Illustrations (as above) reduced in Lithography. Imp. 16mo.
10s. 6d.

HALF-HOUR LECTURES on the HISTORY and PRACTICE of the
FINE and ORNAMENTAL ARTS. By WILLIAM B. SCOTT. Third Edition,
with 50 Woodcuts. Crown 8vo. 8s. 6d.

The THREE CATHEDRALS DEDICATED to ST. PAUL, in LONDON ;
their History from the Foundation of the First Building in the Sixth Century
to the Proposals for the Adornment of the Present Cathedral. By WILLIAM
LONGMAN, F.A.S. With numerous Illustrations. Square crown 8vo. 21s.

IN FAIRYLAND ; Pictures from the Elf-World. By RICHARD
DOYLE. With a Poem by W. ALLINGHAM. With Sixteen Plates, containing
Thirty-six Designs printed in Colours. Second Edition. Folio, price 15s.

The NEW TESTAMENT, illustrated with Wood Engravings after the
Early Masters, chiefly of the Italian School. Crown 4to. 63s. cloth, gilt top ;
or £5. 5s. elegantly bound in morocco.

SACRED and LEGENDARY ART. By MRS. JAMESON. With numerous
Etchings and Engravings on Wood from Early Missals, Mosaics, Illuminated
MSS. and other Original Sources.

LEGENDS of the SAINTS and MARTYRS. Latest Edition, with 19
Etchings and 187 Woodcuts. 2 vols. square crown 8vo. 31s. 6d.

LEGENDS of the MONASTIC ORDERS. Latest Edition, with 11
Etchings and 88 Woodcuts. 1 vol. square crown 8vo. 21s.

LEGENDS of the MADONNA. Latest Edition, with 27 Etchings and
165 Woodcuts. 1 vol. square crown 8vo. 21s.

The HISTORY of OUR LORD, with that of his Types and Precursors.
Completed by Lady EASTLAKE. Latest Edition, with 31 Etchings and
281 Woodcuts. 2 vols. square crown 8vo. 42s.

The Useful Arts, Manufactures, &c.

GWILT'S ENCYCLOPÆDIA of ARCHITECTURE, with above 1,600
Engravings on Wood. New Edition, revised and enlarged by WYATT
PAPWORTH. 8vo. 52s. 6d.

HINTS on HOUSEHOLD TASTE in FURNITURE, UPHOLSTERY,
and other Details. By CHARLES L. EASTLAKE, Architect. Third Edition,
with about 90 Illustrations. Square crown 8vo. 14s.

INDUSTRIAL CHEMISTRY ; a Manual for Manufacturers and for
use in Colleges or Technical Schools. Being a Translation of Professors Stohmann
and Engler's German Edition of PAYEN's *Précis de Chimie Industrielle*, by Dr.
J. D. BARRY. Edited and supplemented by B. H. PAUL, Ph.D. 8vo. with Plates
and Woodcuts. *[In the press.*

URE'S DICTIONARY of ARTS, MANUFACTURES, and MINES.
Seventh Edition, rewritten and enlarged by ROBERT HUNT, F.R.S. assisted by numerous Contributors eminent in Science and the Arts, and familiar with Manufactures. With above 2,100 Woodcuts. 3 vols. medium 8vo. £5 5s.

VOL. IV. Supplementary, completing all the Departments of the Dictionary to the year 1877, is preparing for publication.

HANDBOOK of PRACTICAL TELEGRAPHY. By R. S. CULLEY, Memb. Inst. C.E. Engineer-in-Chief of Telegraphs to the Post Office. Sixth Edition, with 144 Woodcuts and 5 Plates. 8vo. price 16s.

ENCYCLOPÆDIA of CIVIL ENGINEERING, Historical, Theoretical, and Practical. By E. CRESY, C.E. With above 3,000 Woodcuts. 8vo. 42s.

The AMATEUR MECHANIC'S PRACTICAL HANDBOOK; describing the different Tools required in the Workshop, the uses of them, and how to use them, with examples of different kinds of work, &c. with full Descriptions and Drawings. By A. H. G. HOBSON. With 33 Woodcuts. Crown 8vo. 2s. 6d.

The ENGINEER'S VALUING ASSISTANT. By H. D. HOSKOLD, Civil and Mining Engineer, 16 years Mining Engineer to the Dean Forest Iron Company. 8vo.   [In the press.

The WHITWORTH MEASURING MACHINE; including Descriptions of the Surface Plates, Gauges, and other Measuring Instruments made by Sir JOSEPH WHITWORTH, Bart. By T. M. GOODEVE, M.A. and C. P. B. SHELLEY, C.E. Fcp. 4to. with 4 Plates and 44 Woodcuts.   [Nearly ready.

TREATISE on MILLS and MILLWORK. By Sir W. FAIRBAIRN, Bart. F.R.S. New Edition, with 18 Plates and 322 Woodcuts, 2 vols. 8vo. 32s.

USEFUL INFORMATION for ENGINEERS. By Sir W. FAIRBAIRN, Bart. F.R.S. Revised Edition, with Illustrations. 3 vols. crown 8vo. price 31s. 6d.

The APPLICATION of CAST and WROUGHT IRON to Building Purposes. By Sir W. FAIRBAIRN, Bart. F.R.S. Fourth Edition, enlarged; with 6 Plates and 118 Woodcuts. 8vo. price 16s.

The THEORY of STRAINS in GIRDERS and similar Structures, with Observations on the application of Theory to Practice, and Tables of the Strength and other Properties of Materials. By BINDON B. STONEY, M.A. M. Inst. C.E. Second Edition, royal 8vo. with 5 Plates and 123 Woodcuts, 36s.

A TREATISE on the STEAM ENGINE, in its various Applications to Mines, Mills, Steam Navigation, Railways, and Agriculture. By J. BOURNE, C.E. With Portrait, 37 Plates, and 546 Woodcuts. 4to. 42s.

CATECHISM of the STEAM ENGINE, in its various Applications to Mines, Mills, Steam Navigation, Railways, and Agriculture. By the same Author. With 89 Woodcuts. Fcp. 8vo. 6s.

HANDBOOK of the STEAM ENGINE. By the same Author, forming a KEY to the Catechism of the Steam Engine, with 67 Woodcuts. Fcp. 9s.

BOURNE'S RECENT IMPROVEMENTS in the STEAM ENGINE in its various applications to Mines, Mills, Steam Navigation, Railways, and Agriculture. By JOHN BOURNE, C.E. With 124 Woodcuts. Fcp. 8vo. 6s.

LATHES and TURNING, Simple, Mechanical, and Ornamental. By
W. HENRY NORTHCOTT. Second Edition, with 338 Illustrations. 8vo. 18s.

PRACTICAL TREATISE on METALLURGY, adapted from the last
German Edition of Professor KERL's *Metallurgy* by W. CROOKES, F.R.S. &c.
and E. BÖHRIG, Ph.D. M.E. With 625 Woodcuts. 3 vols. 8vo. price £4 19s.

MITCHELL'S MANUAL of PRACTICAL ASSAYING. Fourth Edi-
tion, for the most part rewritten, with all the recent Discoveries incorporated,
by W. CROOKES, F.R.S. With 199 Woodcuts. 8vo. 31s. 6d.

LOUDON'S ENCYCLOPÆDIA of AGRICULTURE: comprising the
Laying-out, Improvement, and Management of Landed Property, and the Cultivation and Economy of Agricultural Produce. With 1,100 Woodcuts. 8vo. 21s.

LOUDON'S ENCYCLOPÆDIA of GARDENING: comprising the
Theory and Practice of Horticulture, Floriculture, Arboriculture, and Landscape
Gardening. With 1,000 Woodcuts. 8vo. 21s.

Religious and *Moral Works.*

**CHRISTIAN LIFE, its COURSE, its HINDRANCES, and its
HELPS**; Sermons preached mostly in the Chapel of Rugby School. By the
late THOMAS ARNOLD, D.D. 8vo. 7s. 6d.

CHRISTIAN LIFE, its HOPES, its FEARS, and its CLOSE;
Sermons preached mostly in the Chapel of Rugby School. By the late
THOMAS ARNOLD, D.D. 8vo. 7s. 6d.

SERMONS chiefly on the INTERPRETATION of SCRIPTURE.
By the late THOMAS ARNOLD, D.D. 8vo. price 7s. 6d.

SERMONS preached in the Chapel of Rugby School; with an Address
before Confirmation. By the late THOMAS ARNOLD, D.D. Fcp. 8vo. 3s. 6d.

THREE ESSAYS on RELIGION: Nature; the Utility of Religion;
Theism. By JOHN STUART MILL. 8vo. price 10s. 6d.

INTRODUCTION to the SCIENCE of RELIGION. Four Lectures
delivered at the Royal Institution; with Two Essays on False Analogies and
the Philosophy of Mythology. By F. MAX MÜLLER, M.A. Crown 8vo. 10s. 6d.

SUPERNATURAL RELIGION; an Inquiry into the Reality of Divine
Revelation. Sixth Edition. 2 vols. 8vo. 24s.

BEHIND the VEIL; an Outline of Bible Metaphysics compared
with Ancient and Modern Thought. By the Rev. T. GRIFFITH, M.A. Pre-
bendary of St. Paul's. 8vo. 10s. 6d.

The TRIDENT, the CRESCENT, and the CROSS; a View of the
Religious History of India during the Hindu, Buddhist, Mohammedan, and
Christian Periods. By the Rev. J. VAUGHAN, Nineteen Years a Missionary of
the C.M.S. in Calcutta. 8vo. 9s. 6d.

The **PRIMITIVE** and **CATHOLIC FAITH** in Relation to the Church of England. By the Rev. B. W. SAVILE, M.A. 8vo. price 7*s.*

SYNONYMS of the **OLD TESTAMENT**, their **BEARING** on **CHRISTIAN FAITH** and **PRACTICE.** By the Rev. R. B. GIRDLESTONE, M.A. 8vo. 15*s.*

An **INTRODUCTION** to the **THEOLOGY** of the **CHURCH** of ENGLAND, in an Exposition of the Thirty-nine Articles. By the Rev. T. P. BOULTBEE, LL.D. Fcp. 8vo. price 6*s.*

An **EXPOSITION** of the **39 ARTICLES**, Historical and Doctrinal. By E. HAROLD BROWNE, D.D. Lord Bishop of Winchester. 8vo. 16*s.*

The **LIFE** and **LETTERS** of **ST. PAUL**, including a New English Translation of the Epistles. By the Rev. W. J. CONYBEARE, M.A. and the Very Rev. JOHN SAUL HOWSON, D.D. Dean of Chester. Copiously illustrated with Landscape Views, Maps, Plans, Charts, Coins, and Vignettes.

    Library Edition, with all the Original Illustrations, Maps, Landscapes on Steel, Woodcuts, &c. 2 vols. 4to. 42*s.*

    Intermediate Edition, with a Selection of Maps, Plates, and Woodcuts. 2 vols. square crown 8vo. 21*s.*

    Student's Edition, revised and condensed, with 46 Illustrations and Maps. 1 vol. crown 8vo. price 9*s.*

HISTORY of the **REFORMATION** in **EUROPE** in the **TIME** of CALVIN. By the Rev. J. H. MERLE D'AUBIGNÉ, D.D. Translated by W. L. R. CATES. (In Eight Volumes.) 7 vols. 8vo. price £5. 11*s.*

VOL. VIII. translated by W. L. R. CATES, completing the English Edition of the Rev. Dr. D'AUBIGNÉ's History of the Reformation in the time of CALVIN, is in the press.

NEW TESTAMENT COMMENTARIES. By the Rev. W. A. O'CONOR. B.A. Rector of St. Simon and St. Jude, Manchester. Crown 8vo.

    Epistle to the Romans, price 3*s.* 6*d.*

    Epistle to the Hebrews, 4*s.* 6*d.*

    St. John's Gospel, 10*s.* 6*d.*

A **CRITICAL** and **GRAMMATICAL COMMENTARY** on **ST. PAUL'S** Epistles. By C. J. ELLICOTT. D.D. Lord Bishop of Gloucester and Bristol. 8vo.

GALATIANS, Fourth Edition, 8*s.* 6*d.*

EPHESIANS, Fourth Edition, 8*s.* 6*d.*

PASTORAL EPISTLES, Fourth Edition, 10*s.* 6*d.*

PHILIPPIANS, COLOSSIANS, and PHILEMON, Third Edition, 10*s.* 6*d.*

THESSALONIANS, Third Edition, 7*s.* 6*d.*

HISTORICAL LECTURES on the **LIFE** of **OUR LORD.** By C. J. ELLICOTT, D.D. Bishop of Gloucester and Bristol. Sixth Edition. 8vo. 12*s.*

EVIDENCE of the TRUTH of the CHRISTIAN RELIGION derived from the Literal Fulfilment of Prophecy. By ALEXANDER KEITH, D.D. 37th Edition, with Plates, in square 8vo. 12s. 6d.; 39th Edition, in post 8vo. 6s.

HISTORY of ISRAEL. By H. EWALD, late Professor of the Univ. of Göttingen. Translated by J. E. CARPENTER, M.A., with a Preface by RUSSELL MARTINEAU, M.A. 5 vols. 8vo. 63s.

The ANTIQUITIES of ISRAEL. By HEINRICH EWALD, late Professor of the University of Göttingen. Translated from the German by HENRY SHAEN SOLLY, M.A. 8vo. price 12s. 6d.

The PROPHETS and PROPHECY of ISRAEL; An Historical and Critical Inquiry. By Dr. A. KUENEN, Prof. of Theol. in the Univ. of Leyden. Translated from the Dutch by the Rev. A. MILROY, M.A. with an Introduction by J. MUIR, D.C.L. 8vo. 21s.

MYTHOLOGY among the HEBREWS, its Historical Development; Researches bearing on the Science of Mythology and the History of Religion. By IGNAZ GOLDZIHER, Ph.D. Member of the Hungarian Academy of Sciences. Translated by RUSSELL MARTINEAU, M.A. 8vo. 16s.

The TREASURY of BIBLE KNOWLEDGE; being a Dictionary of the Books, Persons, Places, Events, and other matters of which mention is made in Holy Scripture. By Rev. J. AYRE, M.A. With Maps, 16 Plates, and numerous Woodcuts. Fcp. 8vo. price 6s. cloth, or 10s. 6d. neatly bound in calf.

LECTURES on the PENTATEUCH and the MOABITE STONE. By the Right Rev. J. W. COLENSO, D.D. Bishop of Natal. 8vo. 12s.

The PENTATEUCH and BOOK of JOSHUA CRITICALLY EXAMINED. By the Right Rev. J. W. COLENSO, D.D. Bishop of Natal. Crown 8vo. 6s.

An INTRODUCTION to the STUDY of the NEW TESTAMENT, Critical, Exegetical, and Theological. By the Rev. S. DAVIDSON, D.D. LL.D. 2 vols. 8vo. price 30s.

SOME QUESTIONS of the DAY. By the Author of 'Amy Herbert.' Crown 8vo. price 2s. 6d.

THOUGHTS for the AGE. By the Author of 'Amy Herbert,' &c. Fcp. 8vo, price 3s. 6d.

PASSING THOUGHTS on RELIGION. By ELIZABETH M. SEWELL. Fcp. 8vo. 3s. 6d.

SELF-EXAMINATION before CONFIRMATION. By ELIZABETH M. SEWELL. 32mo. 1s. 6d.

PREPARATION for the HOLY COMMUNION; the Devotions chiefly from the Works of JEREMY TAYLOR. By Miss SEWELL. 32mo. 3s.

LYRA GERMANICA, Hymns translated from the German by Miss C. WINKWORTH. Fcp. 8vo. price 5s.

SPIRITUAL SONGS for the SUNDAYS and HOLIDAYS throughout the Year. By J. S. B. MONSELL, LL.D. Fcp. 8vo. 5s. 18mo. 2s.

HOURS of THOUGHT on SACRED THINGS; a Volume of Sermons. By James Martineau, D.D. LL.D. Crown 8vo. 7*s.* 6*d.*

ENDEAVOURS after the CHRISTIAN LIFE; Discourses. By the Rev. J. Martineau, LL.D. Fifth Edition. Crown 8vo. 7*s.* 6*d.*

HYMNS of PRAISE and PRAYER, collected and edited by the Rev. J. Martineau, LL.D. Crown 8vo. 4*s.* 6*d.* 32mo. 1*s.* 6*d.*

The **TYPES of GENESIS**, briefly considered as revealing the Development of Human Nature. By Andrew Jukes. Third Edition. Crown 8vo. 7*s.* 6*d.*

The **SECOND DEATH and the RESTITUTION of ALL THINGS**; with some Preliminary Remarks on the Nature and Inspiration of Holy Scripture. By Andrew Jukes. Fourth Edition. Crown 8vo. 3*s.* 6*d.*

WHATELY'S INTRODUCTORY LESSONS on the CHRISTIAN Evidences. 18mo. 6*d.*

BISHOP JEREMY TAYLOR'S ENTIRE WORKS. With Life by Bishop Heber. Revised and corrected by the Rev. C. P. Eden. Complete in Ten Volumes, 8vo. cloth, price £5. 5*s.*

Travels, Voyages, &c.

A YEAR in WESTERN FRANCE. By M. Betham-Edwards. With Frontispiece View of the Hôtel de Ville, La Rochelle. Crown 8vo. 10*s.* 6*d.*

JOURNAL of a RESIDENCE in VIENNA and BERLIN during the eventful Winter, 1805-6. By the late Henry Reeve, M.D. Published by his Son. Crown 8vo. price 8*s.* 6*d.*

The **INDIAN ALPS, and How we Crossed them**: being a Narrative of Two Years' Residence in the Eastern Himalayas, and Two Months' Tour into the Interior. By a Lady Pioneer. With Illustrations from Drawings by the Author. Imperial 8vo. 42*s.*

TYROL and the TYROLESE; being an Account of the People and the Land, in their Social, Sporting, and Mountaineering Aspects. By W. A. Baillie Grohman. With numerous Illustrations Crown 8vo. 14*s.*

'The **FROSTY CAUCASUS**;' An Account of a Walk through Part of the Range, and of an Ascent of Elbruz in the Summer of 1874. By F. C. Grove. With Eight Illustrations and a Map. Crown 8vo. 15*s.*

A THOUSAND MILES up the NILE, being a JOURNEY through EGYPT and NUBIA to the SECOND CATARACT By Amelia B. Edwards. With Eighty Illustrations from Drawings by the Author, Two Maps, Plans, Facsimiles, &c. Imperial 8vo. price 42*s.*

OVER the SEA and FAR AWAY; being a Narrative of a Ramble round the World. By Thomas Woodbine Hinchliff, M.A. F.R.G.S. President of the Alpine Club. With 14 full-page Illustrations. Medium 8vo. 21*s.*

THROUGH BOSNIA and the HERZEGOVINA on FOOT during the INSURRECTION; with an Historical Review of Bosnia, and a Glimpse at the Croats, Slavonians, and the Ancient Republic of Ragusa. By A. J. Evans, B.A. F.S.A. Second Edition, with Map and 58 Wood Engravings. 8vo. 18*s.*

DISCOVERIES at EPHESUS, including the Site and Remains of the Great Temple of Diana. By J. T. WOOD, F.S.A. With 27 Lithographic Plates and 42 Engravings on Wood. Imperial 8vo. price 63s.

MEMORIALS of the DISCOVERY and EARLY SETTLEMENT of the BERMUDAS or SOMERS ISLANDS, from 1615 to 1685. Compiled from the Colonial Records and other original sources. By Major-General J. H. LEFROY, R.A. C.B. F.R.S. &c. Governor of the Bermudas. 8vo. with Map.
[In the press.

ITALIAN ALPS; Sketches in the Mountains of Ticino, Lombardy, the Trentino, and Venetia. By DOUGLAS W. FRESHFIELD, Editor of 'The Alpine Journal.' Square crown 8vo. with Maps and Illustrations, price 15s.

The RIFLE and the HOUND in CEYLON. By Sir SAMUEL W. BAKER, M.A. F.R.G.S. With Illustrations. Crown 8vo. 7s. 6d.

EIGHT YEARS in CEYLON. By Sir SAMUEL W. BAKER, M.A. F.R.G.S. With Illustrations Crown 8vo. 7s. 6d.

TWO YEARS IN FIJI, a Descriptive Narrative of a Residence in the Fijian Group of Islands; with some Account of the Fortunes of Foreign Settlers and Colonists up to the Time of the British Annexation. By LITTON FORBES, M.D. F.R.G.S. Crown 8vo. 8s. 6d.

UNTRODDEN PEAKS and UNFREQUENTED VALLEYS; a Midsummer Ramble among the Dolomites. By AMELIA B. EDWARDS. With a Map and 27 Wood Engravings. Medium 8vo. 21s.

The DOLOMITE MOUNTAINS; Excursions through Tyrol, Carinthia, Carniola, and Friuli, 1861-1863. By J. GILBERT and G. C. CHURCHILL, F.R.G.S. With numerous Illustrations. Square crown 8vo. 21s.

The ALPINE CLUB MAP of SWITZERLAND, with parts of the Neighbouring Countries, on the Scale of Four Miles to an Inch. Edited by R. C. NICHOLS, F.S.A. F.R.G.S. In Four Sheets, price 42s. or mounted in a case, 52s. 6d. Each Sheet may be had separately, price 12s. or mounted in a case, 15s.

MAP of the CHAIN of MONT BLANC, from an Actual Survey in 1863-1864. By ADAMS-REILLY, F.R.G.S. M.A.C. Published under the Authority of the Alpine Club. In Chromolithography on extra stout drawing-paper 28in. × 17in. price 10s. or mounted on canvas in a folding case, 12s. 6d.

HOW to SEE NORWAY. By Captain J. R. CAMPBELL. With Map and 5 Woodcuts. Fcp. 8vo. price 5s.

GUIDE to the PYRENEES, for the use of Mountaineers. By CHARLES PACKE. With Map and Illustrations. Crown 8vo. 7s. 6d.

The ALPINE GUIDE. By JOHN BALL, M.R.I.A. late President of the Alpine Club. 3 vols. post 8vo. Thoroughly Revised Editions, with Maps and Illustrations:—I. *Western Alps,* 6s. 6d. II. *Central Alps,* 7s. 6d. III. *Eastern Alps,* 10s. 6d. Or in Ten Parts, price 2s. 6d. each.

INTRODUCTION on ALPINE TRAVELLING in GENERAL, and on the Geology of the Alps, price 1s. Each of the Three Volumes or Parts of the *Alpine Guide* may be had with this INTRODUCTION prefixed, price 1s. extra.

Works of *Fiction*.

The ATELIER du LYS; or, an Art-Student in the Reign of Terror. By the Author of 'Mademoiselle Mori' Third Edition. 1 vol. crown 8vo. 6*s.*

NOVELS and TALES. By the Right Hon. the EARL of BEACONS-FIELD. Cabinet Edition, complete in Ten Volumes, crown 8vo. price £3.

LOTHAIR, 6*s.*	HENRIETTA TEMPLE, 6*s.*
CONINGSBY, 6*s.*	CONTARINI FLEMING, &c. 6*s.*
SYBIL, 6*s.*	ALROY, IXION, &c. 6*s.*
TANCRED, 6*s.*	The YOUNG DUKE, &c. 6*s.*
VENETIA, 6*s.*	VIVIAN GREY 6*s.*

CABINET EDITION of STORIES and TALES by Miss SEWELL:—

AMY HERBERT, 2*s.* 6*d.*	IVORS, 2*s.* 6*d.*
GERTRUDE, 2*s.* 6*d.*	KATHARINE ASHTON, 2*s.* 6*d*
The EARL'S DAUGHTER, 2*s.* 6*d.*	MARGARET PERCIVAL, 3*s.* 6*d.*
EXPERIENCE *of* LIFE, 2*s.* 6*d.*	LANETON PARSONAGE, 3*s.* 6*d.*
CLEVE HALL, 2*s.* 6*d.*	URSULA, 3*s.* 6*d.*

BECKER'S GALLUS; or, Roman Scenes of the Time of Augustus: with Notes and Excursuses. Post 8vo. 7*s.* 6*d.*

BECKER'S CHARICLES; a Tale illustrative of Private Life among the Ancient Greeks: with Notes and Excursuses. Post 8vo. 7*s.* 6*d.*

HIGGLEDY-PIGGLEDY; or, Stories for Everybody and Everybody's Children. By the Right Hon. E. M. KNATCHBULL-HUGESSEN, M.P. With Nine Illustrations from Designs by R. Doyle. Crown 8vo. 6*s.*

WHISPERS from FAIRYLAND. By the Right Hon. E. H. KNATCH-BULL-HUGESSEN, M.P. With Nine Illustrations. Crown 8vo. 6*s.*

The MODERN NOVELIST'S LIBRARY. Each Work, in crown 8vo. complete in a Single Volume :—

LOTHAIR. By the Right Hon. the EARL of BEACONSFIELD. Price 2*s.* boards, or 2*s.* 6*d.* cloth.

ATHERSTONE PRIORY, 2*s.* boards ; 2*s.* 6*d.* cloth.

BRAMLEY-MOORE'S SIX SISTERS *of the* VALLEYS, 2*s.* boards ; 2*s.* 6*d.* cloth.

The BURGOMASTER'S FAMILY, 2*s.* boards ; 2*s.* 6*d.* cloth.

ELSA, a Tale of the Tyrolean Alps. Translated from the German of WILHELMINE VON HILLERN by Lady WALLACE. 2*s.* boards ; 2*s.* 6*d.* cloth.

MADEMOISELLE MORI, 2*s.* boards ; 2*s.* 6*d.* cloth.

MELVILLE'S DIGBY GRAND, 2*s.* boards ; 2*s.* 6*d.* cloth.

————— GLADIATORS, 2*s* boards ; 2*s.* 6*d.* cloth.

————— GOOD FOR NOTHING, 2*s.* boards ; 2*s.* 6*d.* cloth.

————— HOLMBY HOUSE, 2*s.* boards ; 2*s.* 6*d.* cloth.

————— INTERPRETER, 2*s.* boards ; 2*s.* 6*d.* cloth.

————— KATE COVENTRY, 2*s.* boards ; 2*s.* 6*d.* cloth.

————— QUEEN'S MARIES, 2*s.* boards ; 2*s.* 6*d.* cloth.

————— GENERAL BOUNCE, 2*s.* boards ; 2*s.* 6*d.* cloth.

TROLLOPE'S WARDEN, 2*s.* boards ; 2*s.* 6*d.* cloth.

—————BARCHESTER TOWERS, 2*s.* boards ; 2*s.* 6*d.* cloth.

UNAWARES, a Story of an old French Town, 2*s.* boards. ; 2*s.* 6*d.* cloth.

Poetry and *The Drama.*

POEMS. By William B. Scott. With 17 Etchings by L. A. Tadema and W. B. Scott. Crown 8vo. 15*s*.

MOORE'S IRISH MELODIES, with 161 Steel Plates from Original Drawings by D. Maclise, R.A. and the whole of the Text engraved on the same Plates Super-royal 8vo. 21*s*.

MOORE'S LALLA ROOKH, an Oriental Romance, Tenniel's Edition, with 68 Illustrations from Original Drawings, engraved on Wood by G. Pearson and other Artists. Fcp. 4to. price 21*s*.

SOUTHEY'S POETICAL WORKS, with the Author's last Corrections and copyright Additions. Medium 8vo. with Portrait and Vignette, 14*s*.

LAYS of ANCIENT ROME; with IVRY and the ARMADA. By the Right Hon. Lord Macaulay. 16mo. with Vignettes, 3*s*. 6*d*.

LORD MACAULAY'S LAYS of ANCIENT ROME. With 90 Original tions, chiefly after the Antique, engraved on Wood from Drawings by G. Scharf. Fcp. 4to. 21*s*.

MINIATURE EDITION of LORD MACAULAY'S LAYS of ANCIENT ROME, with the Illustrations (as above) reduced in Lithography. Imperial 16mo. price 10*s*. 6*d*.

The ÆNEID of VIRGIL translated into English Verse. By John Conington, M.A. Crown 8vo. 9*s*.

The ILIAD of HOMER, Homometrically translated by C. B. Cayley, Translator of Dante's Comedy, &c. 8vo. 12*s*. 6*d*.

HORATII OPERA. Library Edition, with Marginal References and English Notes. Edited by the Rev. J. E. Yonge, M.A. 8vo. 21*s*.

The LYCIDAS and EPITAPHIUM DAMONIS of MILTON. Edited, with Notes and Introduction, by C. S. Jerram, M.A. Crown 8vo. 2*s*. 6*d*.

BEOWULF, a Heroic Poem of the Eighth Century (Anglo-Saxon Text and English Translation), with Introduction, Notes, and Appendix. By Thomas Arnold, M.A. Univ. Coll. Oxford. 8vo. 12*s*.

BOWDLER'S FAMILY SHAKSPEARE, cheaper Genuine Editions. Medium 8vo. large type, with 36 Woodcuts, price 14*s*. Cabinet Edition, with the same Illustrations, 6 vols. fcp. 8vo. price 21*s*.

POEMS. By Jean Ingelow. 2 vols. fcp. 8vo. price 10*s*.

> **First Series,** containing 'Divided,' 'The Star's Monument,' &c. Sixteenth Thousand. Fcp. 8vo. price 5*s*.

> **Second Series,** 'A Story of Doom,' 'Gladys and her Island,' &c. Fifth Thousand. Fcp. 8vo. price 5*s*.

POEMS by Jean Ingelow. First Series, with nearly 100 Illustrations, engraved on Wood by Dalziel Brothers. Fcp. 4to. 21*s*.

Rural Sports, &c.

DOWN the ROAD; or, Reminiscences of a Gentleman Coachman. By C. T. S. BIRCH REYNARDSON. Second Edition, with Twelve Coloured Illustrations from Paintings by H. Alken. Medium 8vo. 21*s.*

ANNALS of the ROAD; or, Notes on Mail and Stage Coaching in Great Britain. By CAPTAIN MALET, 18th Hussars. To which are added, Essays on the Road, by NIMROD. With 3 Woodcuts and 10 Coloured Illustrations. Medium 8vo. 21*s.*

ENCYCLOPÆDIA of RURAL SPORTS; a complete Account, Historical, Practical, and Descriptive, of Hunting, Shooting, Fishing, Racing, and all other Rural and Athletic Sports and Pastimes. By D. P. BLAINE. With above 600 Woodcuts (20 from Designs by JOHN LEECH). 8vo. 21*s.*

The FLY-FISHER'S ENTOMOLOGY. By ALFRED RONALDS. With coloured Representations of the Natural and Artificial Insect. Sixth Edition, with 20 coloured Plates. 8vo. 14*s.*

A BOOK on ANGLING; a complete Treatise on the Art of Angling in every branch. By FRANCIS FRANCIS. New Edition, with Portrait and 15 other Plates, plain and coloured. Post 8vo. 15*s.*

WILCOCKS'S SEA-FISHERMAN; comprising the Chief Methods of Hook and Line Fishing, a Glance at Nets, and Remarks on Boats and Boating. New Edition, with 80 Woodcuts. Post 8vo. 12*s.* 6*d.*

HORSES and STABLES. By Colonel F. FITZWYGRAM, XV. the King's Hussars. With Twenty-four Plates of Illustrations, containing very numerous Figures engraved on Wood. 8vo. 10*s.* 6*d.*

The HORSE'S FOOT, and HOW to KEEP it SOUND. By W. MILES. With Illustrations. Imperial 8vo. 12*s.* 6*d.*

A PLAIN TREATISE on HORSE-SHOEING. By W. MILES. Post 8vo. with Illustrations, 2*s.* 6*d.*

STABLES and STABLE-FITTINGS. By W. MILES. Imp. 8vo. with 13 Plates, 15*s.*

REMARKS on HORSES' TEETH, addressed to Purchasers. By W. MILES. Post 8vo. 1*s.* 6*d.*

The HORSE; with a Treatise on Draught. By WILLIAM YOUATT. 8vo. with numerous Woodcuts, 12*s.* 6*d.*

The DOG. By WILLIAM YOUATT. 8vo. with numerous Woodcuts, 6*s.*

The DOG in HEALTH and DISEASE. By STONEHENGE. With 70 Wood Engravings. Square crown 8vo. 7*s.* 6*d.*

The GREYHOUND. By STONEHENGE. Revised Edition, with 25 Portraits of Greyhounds. Square crown 8vo. 15*s.*

The OX; his Diseases and their Treatment: with an Essay on Parturition in the Cow. By J. R. DOBSON. Crown 8vo. with Illustrations, 7*s.* 6*d.*

Works of *Utility* and *General Information.*

The THEORY and PRACTICE of BANKING. By H. D. MACLEOD,
M.A. Barrister-at-Law. Third Edition. 2 vols. 8vo. 26*s.*

The ELEMENTS of BANKING. By HENRY DUNNING MACLEOD,
Esq. M.A. Barrister-at-Law. Crown 8vo. 7*s.* 6*d.*

M'CULLOCH'S DICTIONARY, Practical, Theoretical, and Historical,
of Commerce and Commercial Navigation. New and revised Edition. 8vo. 63*s.*
Second Supplement, price 3*s.* 6*d.*

The CABINET LAWYER; a Popular Digest of the Laws of England,
Civil, Criminal, and Constitutional: intended for Practical Use and General
Information. Twenty-fifth Edition. Fcp. 8vo. price 9*s.*

PEWTNER'S COMPREHENSIVE SPECIFIER; a Guide to the
Practical Specification of every kind of Building-Artificers' Work, with Forms
of Conditions and Agreements. Edited by W. YOUNG. Crown 8vo. 6*s.*

WILLICH'S POPULAR TABLES for ascertaining according to the
Carlisle Table of Mortality the Value of Lifehold, Leasehold, and Church Property,
Renewal Fines, Reversions, &c.; also Interest, Legacy, Succession Duty, and
various other useful Tables. Eighth Edition. Post 8vo. 10*s.*

HINTS to MOTHERS on the MANAGEMENT of their HEALTH
during the Period of Pregnancy and in the Lying-in Room. By the late
THOMAS BULL, M.D. New Edition, revised and improved. Fcp. 8vo. 2*s.* 6*d.*

The MATERNAL MANAGEMENT of CHILDREN in HEALTH and
Disease. By the late THOMAS BULL, M.D. New Edition, revised and improved.
Fcp. 8vo. 2*s.* 6*d.*

The THEORY of the MODERN SCIENTIFIC GAME of WHIST.
By WILLIAM POLE, F.R.S. Eighth Edition, enlarged. Fcp. 8vo. 2*s.* 6*d.*

The CORRECT CARD; or, How to Play at Whist: a Whist Catechism.
By Captain A. CAMPBELL-WALKER, F.R.G.S. late 79th Highlanders; Author of
'The Rifle, its Theory and Practice.' New Edition. 32mo. 2*s.* 6*d.*

CHESS OPENINGS. By F. W. LONGMAN, Balliol College, Oxford.
Second Edition revised. Fcp. 8vo. 2*s.* 6*d.*

THREE HUNDRED ORIGINAL CHESS PROBLEMS and STUDIES.
By JAMES PIERCE, M.A. and W. T. PIERCE. With numerous Diagrams. Square
fcp. 8vo. 7*s.* 6*d.* SUPPLEMENT, price 2*s.* 6*d.*

A PRACTICAL TREATISE on BREWING; with Formulæ for Public Brewers, and Instructions for Private Families. By W. BLACK. 8vo. price 10*s.* 6*d.*

MODERN COOKERY for PRIVATE FAMILIES, reduced to a System of Easy Practice in a Series of carefully-tested Receipts. By ELIZA ACTON. Newly revised and enlarged; with 8 Plates and 150 Woodcuts. Fcp. 8vo. 6*s.*

MAUNDER'S TREASURY of KNOWLEDGE and LIBRARY of Reference; comprising an English Dictionary and Grammar, Universal Gazetteer, Classical Dictionary, Chronology, Law Dictionary, a synopsis of the Peerage, useful Tables, &c. Revised Edition. Fcp. 8vo. 6*s.* cloth, or 10*s.* 6*d.* calf.

MAUNDER'S BIOGRAPHICAL TREASURY. Latest Edition, reconstructed, and partly re-written, with above 1,600 additional Memoirs, by W. L. R. CATES. Fcp. 8vo. 6*s.*

MAUNDER'S SCIENTIFIC and LITERARY TREASURY; a Popular Encyclopædia of Science, Literature, and Art. Latest Edition, in part re-written, with above 1,000 new articles, by J. Y. JOHNSON. Fcp. 8vo. 6*s.*

MAUNDER'S TREASURY of GEOGRAPHY, Physical, Historical, Descriptive, and Political. Edited by W. HUGHES, F.R.G.S. With 7 Maps and 16 Plates. Fcp. 8vo. 6*s.*

MAUNDER'S HISTORICAL TREASURY; General Introductory Outlines of Universal History, and a Series of Separate Histories. Revised by the Rev. G. W. COX, M.A. Fcp. 8vo. 6*s.*

MAUNDER'S TREASURY of NATURAL HISTORY, or Popular Dictionary of Birds, Beasts, Fishes, Reptiles, Insects, and Creeping Things. With above 900 Woodcuts. Fcp. 8vo. price 6*s.* cloth.

MAUNDER'S TREASURY of BOTANY, or Popular Dictionary of the Vegetable Kingdom; including a Glossary of Botanical Terms. Edited by J. LINDLEY, F.R.S. and T. MOORE, F.L.S. assisted by eminent Contributors. With 274 Woodcuts and 20 Steel Plates. Two Parts, fcp. 8vo. 12*s.* cloth.

MAUNDER'S TREASURY of BIBLE KNOWLEDGE; being a Dictionary of the Books, Persons, Places, Events, and other Matters of which mention is made in Holy Scripture. Edited by the Rev. J. AYRE, M.A. With Maps, 16 Plates, and numerous Woodcuts. Fcp. 8vo. price 6*s.* cloth.

INDEX.

Spottiswoode & Co., Printers, New-street Square, London.